THE STORY OF RADIO
TO 5G WIRELESS

"To see how far we've come, you must know how it began."

GEORGE J. WHALEN
NY9A

First edition copyright © George J. Whalen 2021

This book has been written for entertainment and general information. It is copyrighted (other than the FCC article in Addendum A) and is the intellectual property of the author.* All rights are reserved. Reviewers may fairly quote excerpts, with attribution. Other forms of copying are not legal unless written permission is given.

This is a book of history stories. It is not a technical consultation, nor a textbook, nor an academic or market study. The author specifically disclaims expertise in telephone and network technology. Many sources have been consulted. All were believed knowledgeable. But the reader is cautioned that much of history is opinion and hearsay. So, what results when many sources are consulted can be less than factual. As the book spans 188 years, there is no way to know. The author has done his best to verify the stories to the extent possible. But the author is in no way obliged to defend, explain, alter or justify any content, or to reveal sources other than those listed herein.

Only the Federal Communications Commission (FCC) document in Addendum A is "official," as it is a publication of the U.S. Government. No other guarantees of factual authority or validity of opinions are offered. Readers are advised to do their own confirmatory research before acting on anything read herein. As the 5G project continues to develop, it will likely undergo further changes in timing or extent. Accordingly, any content, descriptions, plans, explanations or opinions expressed herein regarding 5G are liable to change without notice.

No liability whatsoever is assumed by the author, and no warranty of any sort is offered, to any person or entity, for anything, by the author or any person or entity involved with this book in any way. Use of anything from this book, or inferred from its content, is entirely at your own risk. Note also that patents, copyrights and trademarks may apply. I do sincerely apologize if I have inadvertently and unintentionally omitted or incorrectly described anyone or anything herein. I admit to being human and fallible. I offer a heartfelt "thank you" for your understanding. By purchasing and/or reading this work, you hereby agree to and accept the terms of this disclaimer.

* The FCC document included in Addendum A is a work of the U.S. Government, included herein as a courtesy, for the reader's reference and convenience. Its inclusion does not imply any endorsement of this book by any government agency. Foreign copyrights may apply. Excerpts from SAE International and the National Highway Traffic Safety Administration (NHTSA) are also published verbatim herein, clearly identified and used with permission. Questions regarding any of these included works should be directed to the FCC, SAE International or NHTSA, and not to the author.

This Book Is Dedicated To:

Joyce Anne Whalen,
KC2EME
My Wonderful, Brilliant Wife
And Lifelong Partner
In Everything!
(Always)

Contents

List of Sources	vii
Preface	ix
Introduction	xi

1 Let's Begin with Some Questions and a Word about Radio Waves — 1
2 Dr. Mahlon G. Loomis, Inventor of Radio (1868) — 5
3 Dr. Nikola Tesla: An Inventor Far, Far Ahead of His Time — 16
4 What Led to Our Need to Stay in Touch? — 20
5 The Early Inventions (1832–1876) — 22
6 Radio, Tesla and Marconi — 25
7 Radio Gains a Voice (1900) — 31
8 Vacuum Electron Tubes Arrive (1904) — 36
9 One Inventor Gave Us Regeneration, the Oscillator, the Superheterodyne Receiver, and Frequency Modulation (1912 to 1933) — 38
10 Two-Way Voice Radio Communication Comes to Cars (1920s) — 41
11 Mobile Radio-Telephone Service (1946) — 43
12 1G Voice-Only Analog Cellular Car Phone Service Arrives (1947) — 45
13 That's Right! Cellular Was Invented for Your Car … Really! — 46
14 A Special Tribute to Al Gross, Inventor of the "Handie-Talkie" (1938) — 48
15 Spread Spectrum: An Amazing Invention by a Remarkable Woman—Hedy Lamarr (1942) — 50
16 Bell Labs Scientists Invent the Transistor (1947) — 52
17 MOSFETs and Integrated Circuits Open a Digital Door to the Future — 54

18 Moore's Law, Digitalization, and Technology Growth	56
19 Motorola Invents the First Handheld Analog Cellular Phone (1973)	58
20 2G Digital Networking and the Digital Cellular Phone Revolution (1980s)	60
21 IBM Pioneers a Personal PDA-Cellular Phone (1991), the 3G Digital Network Arrives (Early 2000s), and Apple's iPhone Debuts (2007)	63
22 4G Digital Cellular Phone/Data Network Arrives (2013)	66
23 An Overview of Wireless Phone and Network Functions	67
24 The 5G Advanced Cellular Phone/Data Network Starts Up (2020)	72
25 You'll Likely Need a New Phone for 5G	74
26 Autos: In Search of Highway Safety	76
27 5G Is All about BIG Data!	78
28 5G New Radio (NR) Technical Summary	80
29 Satellites, Network Speed and Latency	82
30 Low-Earth-Orbit (LEO) Satellites	84
31 How Long Until We Have "Autonomous Vehicles"?	86
32 Robocar Design Is Still Evolving	90
33 We've Got a Way to Go Before Autonomous Car Technology Converges with Still-Building 5G Network Technology	92
34 No Good Estimate of a Finish Date … Yet	94
35 Estimated Car Electronics Costs As Time Goes By	100
Afterword: Is There a Dark Side to 5G?	101
Epilogue: Here's to Your Future!	105
Appendix A	107
About the Author	115

List of Sources

The author gratefully acknowledges the information, images and content graciously contributed by the following, as well as by other sources not identified, or identified beneath each image or appearance.

The Library of Congress, Joshua Levy, PhD, Historian and Librarian
　　Dagogo Alltraide, Author and Producer
　　Craig Swain, Leesburg, VA
　　Edward A. Sharpe, Archivist
　　Southwest Museum of Engineering, Communication and Computation, Glendale, AZ
　　Congress of the United States
　　U.S. Department of Transportation
　　U.S. Department of Commerce
　　National Highway Traffic Safety Administration
　　Federal Communications Commission
　　Bell Labs Nokia
　　Shape.att.com
　　Ford Motor Company
　　General Motors Corporation
　　Wikipedia
　　Archives and Libraries of Yale U. and Harvard U.
　　Google
　　Institute of Electrical and Electronics Engineers
　　Intel
　　SAE International
　　SpaceX

GEORGE J. WHALEN

Statista
Skyworks
Amazon
Viavi
Virgin
IBM
W4CAE.com
Getty Images

Preface

Technology history is my passion. I love to find and share stories about the people who gave us radio ... "wireless." In these pages, I have named the principals and, where I can, those who assisted them. Unfortunately, there are unknown geniuses in our history whose contributions were rarely or never recorded in the analog, careless past. All helped grow electronics from its landline start in 1832 to its wireless 5G roll-out in 2020. Some greatly advanced the state of the art. Some literally saved people's lives. I think all are heroes. There are many more to cover, which I hope to do, in time.

One of my aims has been, wherever possible, to share things no one's ever told you about these people. Those fascinating, interesting sidelights that take time (and digging) to discover. Because, if these stories *were* recorded, the method used was probably ink on paper (not the most durable medium for a long haul through time). No discovery is ever more precious than finding information that fixes the lines to the past and gives us a clear message about the genius insights of our heroes. Since we are now the beneficiaries of all of their work, I am here sending, in all our names, a bouquet to the past, with heartfelt thanks.

And now, my best wishes to you. I wrote this book for your enjoyment. I wish you a bright future and 73.

George J. Whalen

Introduction

What is "wireless"? And "5G"?
It's the buzz ... Everybody's talking about it.
But ... what is it?
When did it start? Why is it "our Future"?
What can it do? When will it get here?

Wireless and radio are one! And 5G is "the Fifth Generation telecom 'new radio' network." It's killer technology, but it's simple if you take it a bite at a time. But, make no mistake, it's big and underway. It will be the largest, most powerful global digital and data cellular network ever conceived by humankind! It's coming because our need to quickly communicate digital data has galloped far beyond the operational limits of any other network on Earth. It is new, high-speed broadband, capable of handling all the terabytes ... petabytes ... and exabytes of data for the countless data-hungry digital devices of our now-and-future world. Over 52 high technologies are converging in it to make 5G the most advanced ever, multipurpose digital telecom and data cellular network. It will provide the data communication that makes our future world "work." From autonomous cars to "smart" homes, businesses and cities, artificial intelligence and robots, entertainment and travel, healthcare, services, surgery ... and more!

But what will make this possible are the historic intellectual leg-ups provided by the technological inventions and discoveries of geniuses over the centuries of our rich past! Because they pushed out the limits of human knowledge and disrupted what was thought possible, seers and engineers today have been able to converge these technologies and make the sum of their functionality into immense super-solutions.

GEORGE J. WHALEN

This book focuses on the history of wireless (radio) technology in the United States, revealing genius, insight, inventions and discoveries in electronics that have brought us today's state of the art. You'll find things in here that no one's ever told you.

Like what? Well, did you know that one incredible American genius invented radio more than 150 years ago? He actually sent wireless signals nearly eighteen miles, without batteries, using electricity drawn by kites and copper antenna wires out of thin air? Meet him in Chapter 2.

And another genius inventor, a naturalized American citizen, 120 years ago predicted a vest-pocket portable cell phone, a global communications network and self-driving cars? Meet him in Chapter 3.

Other geniuses from here and around the world have also given us brilliant inventions. Their stories await you.

1

Let's Begin with Some Questions and a Word about Radio Waves

Do you know who invented radio?
Did you know that man's day job was "dentistry"?
Do you know why many people call radio antennas "aerials"?
Did you know that the Congress of the United States has introduced resolution after resolution naming that man as the true inventor of radio ... but has never passed any of them?
Would it come as a surprise to you to learn that the cell phone (or smartphone) is not something that was dreamed up by a company named after a piece of fruit?
Did you know it really came from a company that painted its computers blue?
Or that phones that used radio waves were dubbed "wireless" (instead of "radio") because carrier marketers wanted to charge a premium price for an upscale-sounding service?
Or that Silicon Valley began because one of the inventors of the transistor moved there?
The answers are inside.
But, before we get to them, it would be great to spend a little time answering a more basic question about radio waves, since just about everything you'll read about here relies on the ways they are generated, propagate, behave and perform.

What Are Radio Waves?

Radio waves are the lower-frequency part of the *electromagnetic (EM) spectrum*. They are sine waves ranging from a low frequency of about 30 kilohertz (that is, 30,000 cycles per second) up to an ultra-high 30 gigahertz (30 billion cycles per second). A *cycle* is when a point on a wave passes the same point on the next wave, and frequency is how often this occurs, measured against time. So, *frequency* tells us how many cycles a wave completes in a given period. The higher the frequency, the more cycles. Simple enough. We also have *wavelength*: the measurable distance between identical points on two sequential cycles. At 1 megahertz, a wavelength would be 300 meters (~985 feet). At 30 gigahertz, a wavelength would be 1 centimeter (0.4 inch). Wavelength is critical in the design of radio antennas.

To honor the great physicist Heinrich Hertz, who proved the existence of EM waves in 1886, we say "Hertz" instead of "cycles per second." The prefixes kilo-, mega-, or giga- may be used as the frequency increases. Wavelength is symbolized by the Greek letter lambda and is measured in kilometers, meters, centimeters and millimeters as frequency increases.

Radio waves are invisible to us. But a bit higher in the spectrum are the light waves we can feel as heat (infrared) or see (visible). (Radio waves will be the only ones covered in this book, but what you find here can broadly be applied to the entire EM spectrum.)

So, why do we call these waves "electromagnetic"? Just look at this fascinating drawing. As you can see, the EM radio wave has two parts: an *electric field half*, of a given frequency, and a *magnetic field half*, of exactly the same frequency. These two wave-halves are always at a right angle (90 degrees out of phase) with respect to one another, and they move together. The electric half can actually generate the magnetic half in a conductor! And, the magnetic half can generate the electric half in a conductor!

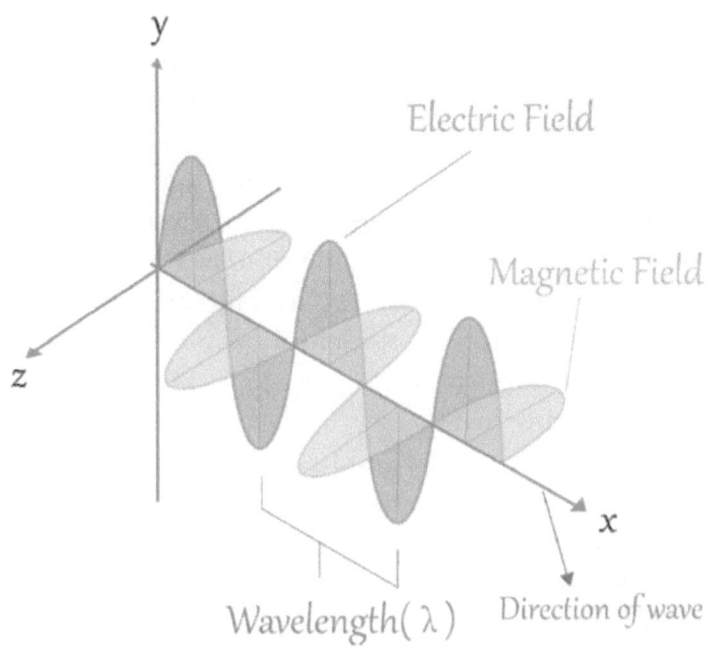

How does a radio wave begin? It starts when *charge carriers* (say, electrons) move up and down in a conductor (such as a wire or an antenna). If the charge carriers are *oscillated* (that is, moved back and forth in the conductor by a reversing-polarity alternating source, such as RF energy) a *near field* is generated about the conductor, radiating electric and magnetic energy at the radio frequency. Once it comes into existence, this radio wave field moves outward at the speed of light (186,000 miles per second, or 300 million meters per second), independent of the energy source that created it. The radio wave can *propagate* over a substantial distance, radiating through a vacuum and most media to become the *far field* at a distance from its origin. A radio wave decreases in strength in proportion to the square of the distance. It is also *polarized* vertically, horizontally, or at any angle in between, by the relative positioning of the transmitting and receiving antennas. This determines the radio wave's direction, and thus the degree of energy (signal) that can be induced into the receiving antenna.

When the radio wave is intercepted by a similarly polarized receiving antenna, its electric and magnetic energy induces a maximum signal in the conductor, with content identical in frequency, duration and modulation. This signal is conducted into the receiver, where it is processed. Cross-polarization decreases the signal.

Radio waves of different frequency, modulation, pulsation and the like can be generated at the very same time by multiple sources and emitted into the medium. These waves can be separated and selected by *tuning* receiving devices to resonantly choose specific-frequency signals out of the panoply of waves arriving at any given time.

For simplicity, this brief "how it works" summary has ignored some of the elegance, richness and beauty in the technical theory of the radio wave. For more information, refer to the ARRL *Handbook for Radio Communications*, as well as the many excellent recorded lectures on EM waves on YouTube. (As this is physics, there will be some math.)

Now, let's meet the inventor of radio.

2

Dr. Mahlon G. Loomis, Inventor of Radio (1868)

Mahlon G. Loomis, aerial telegraphy (radio) inventor

Mahlon Loomis was born in the upstate New York, Fulton County town of Oppenheim, on July 21, 1826, the fourth of nine children. Both his parents were well educated and encouraged reading, study and creative thinking, guiding young Mahlon's curious and inventive mind. His father, Professor Nathan Loomis, a graduate of the Lawrence Scientific School of Harvard, was a teacher, mathematician, writer and, later, special assistant at *The American Ephemeris and National Almanac*, published from Cambridge, Massachusetts. It served seafarers, the U.S. Navy, navigators, astronomers and science-minded readers. Among Mahlon's brothers was George, an inventor who had several patents. Another brother, Joseph, worked in the U.S. Patent Office. And his third brother, Eben, was an astronomer, meteorologist, and nature writer. So, it would be fair to rank this family rather high in terms of its knowledge of science.

The publication Nathan Loomis worked for was an American cousin of the original British *Ephemeris*, a publication instituted by the Crown in England in 1757, well before America's independence. It helped to guide Britain's ships by the stars, making that nation a true naval superpower and global trader in its time. As the United States emerged, its navy and merchant marine had need of reliable charts of our heavens, to safely guide our navigators and tall ships. The elder Loomis had this responsibility, which involved much calculation. Thus, he held the title "Computer and Special Assistant."

The prestigious publication evolved, as did the United States Naval Observatory, which was later relocated from Cambridge to One Observatory Way, Washington, D.C. (Eventually, when the naval observatory relocated again, this became the residence address of the Vice President of the United States.)

The Loomis family moved to Springvale, Virginia, in 1836, when Mahlon was ten. Just twelve years later, he was in Cleveland, Ohio, studying dentistry. Returning to Springvale, he married Ms. Achsie Ashley, and his bride and he soon moved to Washington, where he set up his dental practice. (In the process, he also successfully invented and patented a line of artificial teeth, though his interest in electricity was rising fast.)

By 1860, Mahlon Loomis had delved much more deeply into electricity, spending his off-hours consumed in learning about and experimenting with it. He theorized that the Earth's atmosphere formed a kind of "circuit," girdling the Earth. He felt it was made up of layers of air, and that each layer carried a higher positive electrical charge than the one below it. At ground level, the charge was zero, but it increased to higher and higher voltages with altitude. These thoughts had started with Ben Franklin. But Loomis' thoughts went higher. His idea was that voltage charges increased all the way up to the highest level of the atmosphere (the part we would later call "the ionosphere"). So, he was one of the very first to recognize the existence of atmospheric electricity, which has become a legitimate area of study in physics. (His father and siblings all had exposure to scientific information and may have helped him arrive at his conclusions.)

Let's consider why the Earth's atmosphere would be electrically charged. To begin with, some 40,000 thunderstorms each day rage all over the globe! And it has been shown that cosmic rays and solar radiation continually bombard the upper atmosphere, creating ions that charge it. But Loomis could not have known all this at the time, as these facts were discovered later.

It was not until the mid-twentieth century (1957) that scientist J. Alan Chalmers published his book *Atmospheric Electricity*, describing it this way:

> On an ordinary day over flat desert country, as one goes upward from the surface of the ground, the electric potential increases by about 100 volts per meter. Thus, there is a vertical electric field (E) rising at 100 volts per meter in the air of the atmosphere. The + sign of the E field corresponds to a negative charge on the Earth's surface. So, outdoors the potential at the height of your nose might be +200 volts higher than the potential at your feet.

(The current obtainable from this electric field would be limited by the size of the collector used: microamperes for a pinpoint, but possibly an ampere or more for a large-diameter metal bowl.)

We cannot know whether Mahlon Loomis prematurely grasped this through his own intuition, or whether it came from exposure to his science-oriented family. But the fact remains: somehow, he knew it and used it to conceive of his invention—sending wireless telegraph messages—and it worked! He left us a clear description of his experiment. All that he omitted was the "theory" ... because in 1868, radio and electromagnetic waves were not yet known by physics. Think about that! To really appreciate this man's true insight, realize that he could not possibly have known these things ... because they hadn't yet been discovered.

You see, James Clerk Maxwell, the brilliant mathematician who would, one day, create the complex equations that predicted the existence of electromagnetic waves, was living in Edinburgh, Scotland, when Loomis was performing his aerial telegraph experiment in Virginia, and Maxwell's *Treatise on Electromagnetism* was not printed until 1873. (Hmmm, no help there.)

And Heinrich Hertz, the brilliant German physicist who would actually prove the existence of electromagnetic waves (using electric sparks), did not conduct his famous experiment until 1886, fully eighteen years after Loomis. (Hmmm, no help there, either.)

So, since Loomis could not have had "the theory" from others to express the incredible insights that led him to perform the amazing experiment he did, let's fill in what was missing. (You do not have to read what follows if "techie stuff" gives you a headache. But please believe that what Loomis came up with on his own is startlingly brilliant! If possible, you may want to share this with a friend who is further along in radio technology and have him or her give you an opinion.)

Consider this: Loomis apparently came up with this technical plan on his own.

First, he decided that the ground could serve as one conductor of his telegraph circuit. That had already been done by landline telegraphers, so Loomis' intuition that the ground could be one of his "wires" and also the negative terminal of his electrical source showed real brilliance.

Loomis next intuited that if he tapped into the atmosphere's electric charge at a high enough physical altitude and voltage level, he would have both the positive terminal of his energy source and the "other wire" to complete his circuit.

His final, truly brilliant intuition was that he could use as his "transmitter" and "receiver" two kites, each with a copper wire cord ("aerial") 600 feet long. Each kite was also backed with a lightweight copper screen, connected to the copper wire cord. The kites were to be launched and flown from two high mountains (about 1300 feet) separated by fourteen to eighteen miles. An atmospheric electrical voltage potential as high as hundreds of volts would thus be extracted from the thin air, causing a current flow in the copper wire on the "transmitter kite." This current could be interrupted ("keyed") to create a message of sorts. The moving charge carriers causing the current would generate electromagnetic waves ... radio signals.

Travelling (at the speed of light) each wave would pass through the 600-foot copper wire "aerial" of the second ("receiver") kite, flown from atop the other 1300-foot mountaintop. The wave bursts would induce in this second kite signal voltages resembling the original signal. By connecting the copper wire on the receiving kite through a galvanometer to the ground, Loomis created his "telegraph circuit." So, by opening the connection between the ground and the galvanometer on the transmitting kite, Loomis could start or stop electric current flow and send out his "aerial telegraph" message! The direction of transmission could be reversed simply by re-connecting the ground wires on the galvanometers at the two mountaintop stations. This was radio! (By the way, when Heinrich Hertz got around to discovering electromagnetic waves eighteen years later, he used a small spark-gap transmitter and a handheld receiving ring antenna with a spark gap.)

Thanks to the Southwest Museum of Engineering, Communication and Computation, Glendale, Arizona 85301 (www.smecc.org) for initial information that led me to the work of Dr. Mahlon Loomis. Thanks, too, to Dr. Joshua Levy, historian and librarian at the Library of Congress, for his outstanding service and true dedication to helping people like me get it right.

The Actual Loomis Experiment, as Recorded in His Handwriting, in His Notebook, Now in the Library of Congress

Here, in his own words, Loomis describes the actual witnessed "aerial telegraph" experiment, October 1868:

> From two mountain peaks of the Blue Ridge in Virginia. Which are only about two thousand feet above tide water, two kites were let up—one from each summit—18 miles apart. These kites each had a piece of fine wire copper screen about fifteen inches square attached to the underside and connected also with the (copper) wire six-hundred feet in length which held the kites when they were up. The day was clear and cool in the month of October, with breeze enough to hold the kites firmly at anchor when they were flown. Good connection was made with the ground by laying in a wet place a coil of wire, one end of which was secured to the binding post of a galvanometer. The equipments [sic] and apparatus at both stations were exactly alike. The timepieces of both parties having been set alike, it was arranged that at precisely such an hour and minutes, the galvanometer at one station should be attached, or be in circuit with the ground and kite wires. At the opposite station, the ground wire being already fast to the galvanometer, three separate and deliberate half-minute connections were made with the kite wire and the instrument. This deflected or moved the needle at the other station, with the same vigor and precision as if it had been attached to an ordinary battery. After a lapse of five minutes, as previously

arranged, the same performance was repeated with the same result, until the third time. Then, fifteen minutes precisely were allowed to elapse, during which time the instrument at the first station was put in circuit with both wires while the opposite one was detached from its upper wire, securing arrangements at each station. At the expiration of the fifteen minutes, the signals came into the initial station, a perfect duplicate of those sent from it, as if by previous agreement. And, although no "transmitting key" was made use of nor any "sounder" to voice the signals, yet they were as exact, as precise and distinct as any that ever sped over a wire. A solemn feeling seemed to be impressed upon those who witnessed this little performance, as if some great mystery hovered there around that simpler scene, notwithstanding the results were confidently expected, although the experiment had been continued for two days before the line would "work." And though it continued to transmit signals only about three hours, when from some cause the circuit became inoperative, by the moving away of the upper electric body. Hence it is that high regions must be sought, where disturbing influences cannot invade, where statical energy is spread in a vast, unbroken element, enabling a line to work without interruption or possible failure. No speculation need be indulged as to whether the theory is current, for theory and speculation must stand aside whether they will or not and square themselves with demonstrated truth.

(The nine-page, handwritten text by Mahlon Loomis is his first-person account of the discovery of radio. The original document is in his notebook, in a file carefully preserved by the Library of Congress, in Washington, D.C.)

A patent application was written, and it is said to have been witnessed by individuals from the Smithsonian Institution. U.S. Patent number 129971 was issued to Loomis in 1872. He formed the Loomis

Aerial Telegraphy Co., and its incorporation papers were signed by President Grant. But efforts to raise capital were devastated by bad luck. Times were very hard. Reconstruction after the Civil War dried up all sources of capital. The Congress promised funds, but failed to appropriate them. Then, private investors in Chicago promised to provide Loomis $20,000. But they were suddenly wiped out by the Great Chicago fire! Even harder times followed as the Panic of 1873 destroyed banks. Eighty-nine railroads defaulted on their bonds! No one could invest in anything. So Loomis' invention languished.

Also, since atmospheric electricity is quite changeable, demonstrations sometimes failed and doubts surfaced. Loomis became the butt of cruel jokes from ignorant people. In desperation, he spent his own funds. It cost him his marriage. Yet Loomis remained convinced of the truth of his discovery.

His health declining, he retreated to the farm of his brother, George Loomis. As his life ebbed away, he spoke these last words to his brother:

> I know that I am by some, even many, regarded as a crank—by some perhaps a fool ... but I know that I am right, and if the present generation lives long enough their opinions will be changed ... and their wonder will be that they did not perceive it before. I shall never see it perfected—but it will be, and others will have the honor of the discovery.

Mahlon Loomis lived only sixty years and passed away in 1886, just a short time before Nikola Tesla came to America in 1888. Tesla would take up the torch of wireless telegraphy. Perhaps he also shared some of Mahlon Loomis' ideas about atmospheric electricity. In any event, the early genius and commitment of Mahlon Loomis, the first wireless telegrapher and inventor of radio, surely deserves America's long overdue recognition and admiration. Toward that end, the Congress of the United States has introduced many, many resolutions, each naming Mahlon Loomis "the inventor of radio." Sadly, Congress has also failed to pass any of them and never said why.

A sketch of the successful experiment that demonstrated the phenomenon of wireless transmission by Loomis' "aerial telegraph." (The mountains were Catoctin Ridge and Bears Den.) It is said that an observer from the Smithsonian Institution was a witness.

The U.S. Patent issued to Dr. Loomis is no. 129971. It is reproduced on the following page.

GEORGE J. WHALEN

UNITED STATES PATENT OFFICE.

MAHLON LOOMIS, OF WASHINGTON, DISTRICT OF COLUMBIA.

IMPROVEMENT IN TELEGRAPHING.

Specification forming part of Letters Patent No. **129,971**, dated July 30, 1872.

To all whom it may concern:

Be it known that I, MAHLON LOOMIS, dentist, of Washington, District of Columbia, have invented or discovered a new and Improved Mode of Telegraphing and of Generating Light, Heat, and Motive-Power; and I do hereby declare that the following is a full description thereof.

The nature of my invention or discovery consists, in general terms, of utilizing natural electricity and establishing an electrical current or circuit for telegraphic and other purposes without the aid of wires, artificial batteries, or cables to form such electrical circuit, and yet communicate from one continent of the globe to another.

To enable others skilled in electrical science to make use of my discovery, I will proceed to describe the arrangements and mode of operation.

As in dispensing with the double wire, (which was first used in telegraphing,) and making use of but one, substituting the earth instead of a wire to form one-half the circuit, so I now dispense with both wires, using the earth as one-half the circuit and the continuous electrical element far above the earth's surface for the other part of the circuit. I also dispense with all artificial batteries, but use the free electricity of the atmosphere, co-operating with that of the earth, to supply the electrical dynamic force or current for telegraphing and for other useful purposes, such as light, heat, and motive power.

As atmospheric electricity is found more and more abundant when moisture, clouds, heated currents of air, and other dissipating influences are left below and a greater altitude attained, my plan is to seek as high an elevation as practicable on the tops of high mountains, and thus penetrate or establish electrical connection with the atmospheric stratum or ocean overlying local disturbances. Upon these mountain-tops I erect suitable towers and apparatus to attract the electricity, or, in other words, to disturb the electrical equilibrium, and thus obtain a current of electricity, or shocks or pulsations, which traverse or disturb the positive electrical body of the atmosphere above and between two given points by communicating it to the negative electrical body in the earth below, to form the electrical circuit.

I deem it expedient to use an insulated wire or conductor as forming a part of the local apparatus and for conducting the electricity down to the foot of the mountain, or as far away as may be convenient for a telegraph-office, or to utilize it for other purposes.

I do not claim any new key-board nor any new alphabet or signals; I do not claim any new register or recording instrument; but

What I claim as my invention or discovery, and desire to secure by Letters Patent, is—

The utilization of natural electricity from elevated points by connecting the opposite polarity of the celestial and terrestrial bodies of electricity at different points by suitable conductors, and, for telegraphic purposes, relying upon the disturbance produced in the two electro-opposite bodies (of the earth and atmosphere) by an interruption of the continuity of one of the conductors from the electrical body being indicated upon its opposite or corresponding terminus, and thus producing a circuit or communication between the two without an artificial battery or the further use of wires or cables to connect the co-operating stations.

MAHLON LOOMIS.

Witnesses:
BOYD ELIOT,
C. C. WILSON.

14

Mr. Craig Swain of Leesburg, VA, photographed this official state marker (placed in 1948) in 2007. It has since been reported missing.

3

Dr. Nikola Tesla: An Inventor Far, Far Ahead of His Time

I'm about to share several things no one's ever told you about Tesla. But, first, we'll begin with the fact that he was an incredible wizard to whom we owe abundant thanks for working out our alternating current electrical power system, induction motors, generators, devices and infrastructure. As if that's not enough, he also continued the invention of radio. So, Tesla established TWO big areas of invention on which we have built our technology and economy in modern times.

Dr. Nikola Tesla (1856–1943) was a Serb, born in Smiljan, Croatia, in 1865, and became a naturalized U.S. citizen in 1888. An inventor, genius, futurist,

THE STORY OF RADIO TO 5G WIRELESS

and mechanical and electrical engineer, he is said to have had over 300 patents in 26 countries. Among these patents are devices for use in generation and transmission of alternating current electricity, AC motors, radio (wireless), the remote control, lighting systems, and more. (Photo 1890, the Library of Congress)

It is our good fortune that we have had so many other brilliant, creative scientists, inventors and engineers, over time, who have contributed great, powerful technologies. Now, in the twenty-first century, many of these wonders are converging in the incredible 5G Network.

Of course, you probably know that automobiles were invented way back in the nineteenth century, well over 120 years ago. But did you know that the man pictured above, who lived back then, predicted the concept of self-driving cars? What's more, this very same man also predicted something remarkably like the 5G cellular phone and its global network in "the jazz era" (1926) ... just a few years into the twentieth century.

After World War I (1914 to 1918), electrical innovations in American cars and homes came from some of America's greatest inventors and engineers, employed by industrialists such as Henry Ford, Walter Chrysler, Alfred Sloan and Richard Buick, among others.

Arthur Atwater Kent not only manufactured beautiful early radio equipment fashioned like scientific instruments and high-end furniture, prized for its quality, in the 1920s, but also patented the modern form of the automotive ignition coil. His large plant in North Philadelphia employed 12,000 workers. Out in the Midwest, a little company called Motorola (founded in 1928, in Chicago, by brothers Paul and Joseph Galvin) helped things along, with early radios that could be used in automobiles.

And, consider this: cellular phone technology actually began life in a 1947 Bell Labs U.S. patent on an analog mobile car radiotelephone system by engineering genius W. Rae Young. Fortunately, another brilliant Bell Labs engineer, Jesse Russell, later championed taking it out of the car, making it digital and putting it into people's pockets and purses. His story follows a little later.

What is a bit spooky is that what we are now working toward in the twenty-first century was quite clear to Nikola Tesla in the nineteenth century! Below are two predictions he made long before anyone else:

Tesla predicted autonomous vehicles back in the nineteenth century.
He said:

> As early as 1898, I proposed to representatives of a large manufacturing concern, the construction and public exhibition of an automobile carriage, which, left to itself, would perform a great variety of operations, involving something akin to judgement. (*My Inventions*, Nikola Tesla's autobiography, 1919, page 93)

And then, while being interviewed by *Collier's Weekly* in 1926, for an article about his many accomplishments, Tesla predicted the concepts behind today's cellular "smartphone" and the internet. He described future wireless (radio) devices that would incorporate "television" and "telephone" technologies and work over a global "network" (much like the new 5G network.)

Tesla predicted cellular phones and the internet in 1926.
He said:

> When wireless is perfectly applied to the whole Earth, we shall be able to communicate with one another instantly, irrespective of distance. Not only this, but through television and telephony, we shall see each other as if we were face- to-face, despite the thousands of miles, and the instruments with which we will be able to do this will be amazingly simple, compared to the telephone. A man will be able to carry one in his vest pocket. (From *New Thinking*, book and documentary video by Dagogo Altraide, 2019)

And consider this: just twelve years after Tesla's 1926 prediction, the first *handheld* two-way radio-telephone was invented in 1938! It was a *portable* you could carry with you and use anywhere, created by an American amateur radio operator. His invention went on to become a U.S. Army World War II "secret weapon" that helped rescue downed pilots and save many lives. (More about this later.) And, oh yeah ... that radio is the true ancestor of every personal pocket cellular telephone that has ever been made.

Then, the "internet" arrived.
This remarkable fault-tolerant network came in the 1960s, as ARPANET, a government project, and it grew into its present global form in the 1990s. It takes its name from *Inter*connected and *Net*work. It was what Tesla had in mind for his global "wireless" network. The internet's present form is a complex 750,000-mile matrix of undersea cables and land computer networks that link sites and devices worldwide—in private, public, academic, business and government settings of local to global scope. All are connected by a broad array of electronic, wireless and optical networking technologies. The World Wide Web, created in 1991 by Tim Berners-Lee, a brilliant, celebrated British scientist, is the immeasurably wonderful gift he gave freely to billions of the world's people, who use it for search, contact and information-sharing every day.

So, here's to the inventors, engineers and scientists who've given us the wonderful technologies of today!
Fortunately, history has given us people gifted with the capacity to recognize problems and synthesize ingenious solutions. The main intent of this book is to tell you interesting little-known stories related to these inventors and their inventions, leading up to the 5G network era. To properly cover them all, though, would take an entire shelf of books.

4

What Led to Our Need to Stay in Touch?

In the nineteenth century, travel on land was dreadful—primitive and hard. Few roads existed. Most were rutted. Few were paved. People relied on horses, draft animals and wagons to go places. Most people lived their entire lives within 30 miles of where they were born. Communication was mostly by voice, shout or mail. Travel by ship meant being out of touch and in peril on the high seas. On land, travel by stagecoach, wagon, horse or mule also meant being in peril and out of touch. Once automobiles were invented in the late nineteenth century, the search for new means of communication intensified.

Cars, those ingenious mechanical devices, were first created by Karl Benz, Gottlieb Daimler, Nicolaus Otto, the Duryea brothers and Emile Levassor. They gave humankind mobility. That increased our need to communicate ... and this inspired a new electrical age.

It was Charles F. Kettering's starting motor (1909) and reliable battery ignition (1911), which started, timed and fired the fuel-air mix, that drove the pistons in later General Motors Cadillac cars. (Kettering and GM head Henry Leland were close friends.) Not long after came electric fuel-handling, fuel-injection, lighting, signaling and radio systems to increase safety, communication, performance and enjoyment. Thus, electricity found an early home in automotive invention and technology, giving people new mobility, convenience and freedom to go further.

But we just could not accept being out of touch. We needed communication, like the phones we had at home. So, cars became more electrical and electronic in the twentieth century, bringing mobile radio-telephone service. That evolved into car-borne cellular radio-phones—then blossomed into personal digital cellular phones.

We called them "wireless." Meanwhile, other electronic devices grew, and everything in our lives became "digital." That's why we create and now use immense and growing volumes of data. And more will be needed for the next generation of autonomous cars. Over the past 120 years, cars in the United States have grown in number to our present (2020) fleet of 280 million. Of that number, 3 percent are now battery electric vehicles (EVs); the bulk are internal combustion engine (ICE) vehicles. Our ever-expanding mobility has multiplied our travel and our need to communicate, pushing technology to advance and make it possible.

5

The Early Inventions (1832–1876)

The First "Networks": Landline Telegraphy and Landline Telephone

A number of geniuses deserve to take a deep bow here. Believe it or not, the telegraph served as an inspiration to Alexander Graham Bell, who, indeed, invented the telephone, but was really trying to invent an acoustic telegraph device. Remember, too, that wireless radio-telegraphy, when it came, built upon the foundation that was laid by its land-based predecessors.

Samuel F. B. Morse (left; telegraphy, 1832) and Alexander Graham Bell (right, at the New York Stock Exchange; telephone, 1876)

Morse code was devised by Samuel F. B. Morse for his early telegraph operators, and it has lasted well beyond his messaging system of keys, poles and wires. His code has endured for over 190 years and is still in daily use throughout the world in all forms of telegraphy.

As Bell was working on his acoustic telegraph model one day, he idly plucked a magnetized reed fitted onto a coil in the device. The musical note it sounded was clearly heard in the room, but also in an earphone worn by his assistant, several rooms away. Sound had been communicated over wires by electricity!

Other experiments followed. A rudimentary, cup-shaped microphone was put into the circuit. Bell poured a tiny amount of electrolyte (acid) into it, but spilled a bit of it on his pant leg, burning himself. He cried out in pain, and suddenly the telephone spoke for the very first time: "Mr. Watson ... come here, I want you!"

So, it is certainly right and proper that we first give Samuel Morse, assisted by Charles Jackson, a good round of applause for the telegraph (1832). Then, add a big cheer for A. G. Bell, assisted by Thomas Watson (1876), for the telephone. And add two more, for Antonio Meucci (1854) and Elisha Gray (1876—who filed his patent application just one day after Bell). These inventors all separately created functioning telephones (but because Bell had date priority, he got the U.S. patent).

Carbon Microphone

Not long afterward, the carbon microphone was independently developed (1878) by David E. Hughes (in Great Britain) and by Emile Berliner and Thomas Edison (in the United States). It had a diaphragm that moved in response to sound and proportionally varied the electrical resistance of a chamber filled with tiny carbon granules. Connected to a battery, it could provide a variable flow of current representing speech or other sound. It soon became part of the telephone.

Electromagnetic Waves

Don't forget Heinrich R. Hertz, an eminent physicist and the very first (1886) to conclusively prove the existence of electromagnetic (radio) waves. These had been predicted by genius mathematician James C. Maxwell's earlier equations of electromagnetism. But Dr. Hertz revealed them. The unit of frequency, cycles per second, is named "Hertz" to honor this great scientist's discovery.

6

Radio, Tesla and Marconi

Once and for all, it was Nikola Tesla (not Guglielmo Marconi, as is so often reported) who truly invented a radio-telegraphy system. Tesla held the issued patents. As early as 1892, Tesla had created, and legally patented in the United States, the designs for all his radio-telegraphy apparatus.

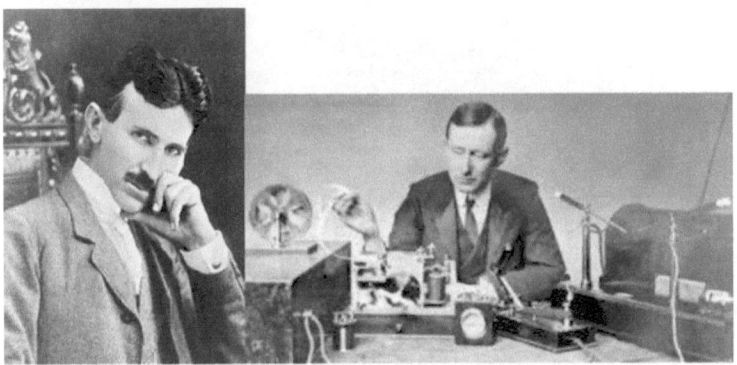

Nikola Tesla (left, 1892) and Guglielmo Marconi (right, 1899)

Though Marconi later applied for patents covering the "embodiments" of equipment he made, he had no lawful right to use Tesla's intellectual property—and he could not even explain how radio worked! But Marconi came from wealth and nobility (he was a marquis). His father was from an old noble Italian family; his mother was Irish, hailing from the Jameson distilling clan.

Marconi was an early believer in radio's potential and simply began following his dream. He also had all the rich and powerful sponsors, the connections and the money he needed.

So, he took and used Tesla's radio inventions. It is said he used seventeen of Tesla's patents to create his long-distance radiotelegraphy system. Tubes did not exist at that time. So, his system used a Tesla coil spark-gap transmitter, which emitted on all frequencies at once.

A compact spark-gap transmitter. This lower-power spark-gap transmitter was made for local or amateur use. Its induction coil is in the enclosure. The gap is adjustable and the spark occurs between the two brass balls on the ends of the adjustment rods atop the insulators. Connections are provided for primary power, antenna and ground. This type of transmitter was popular at the turn of the nineteenth century and is now found mostly in museums.

In 1899, Marconi won a shared Nobel Prize (with Karl F. Braun, inventor of resonant circuit tuning devices and receiving detectors) for advancing "long-distance communication."

Using radio-telegraphy, Marconi had managed to send a message (mostly lost in static, but one letter "S" was heard) 2,100 miles across the Atlantic, from Poldhu, Cornwall, to St. John's, Newfoundland. It was his plan to offer complete radio-telegraphy services for hire, and he organized the Marconi Telegraph Company in London. This company manufactured spark-gap transmitters and receivers, and it also trained men to use the equipment at customer locations and provided installation and service. Thus, Marconi offered a complete solution for those in need of messaging services. He marketed relentlessly.

Masterfully equipping the British royal yacht with radio equipment, he soon had Queen Victoria sending radio messages to the Prince of Wales, her son Albert Edward. This became the fashion of the day! Soon Marconi's company began outfitting British ships with spark-gap radio-telegraph systems and training operators to be rented out to serve on ships, as well as at land stations.

Again and again, Tesla brought patent infringement suits against Marconi over the years, but he could not begin to match the opposition mounted against him by the towering legal, royal and financial resources of Marconi's connections. Tesla's efforts to secure justice were thwarted throughout his lifetime.

In his 1909 speech of acceptance of his shared Nobel Prize, Marconi honestly admitted that he had no idea how radio worked. At the time, the huge, high-powered spark discharges from Tesla coils were the source of his damped-wave radio transmissions. In headphones, they sounded exactly like the hissing electric sparks they were!

So, what befell Tesla was not simply the injustice of his patents being looted, but also the neglect and bad fortune that lack of popular recognition of his invention brought.

Yet, despite this, a measure of credit still must be given to Marconi for having developed the means, the organization and the resources that actually made radio-telegraphy services possible.

The *Titanic* Disaster

It is worth remembering that all of the 706 survivors of the *RMS Titanic* disaster in 1912 were saved because of three heroic Marconi radiomen—Harold Bride, Jack Philips, and Harold Cottam.

Bride and Philips kept sending SOS messages on the doomed *Titanic*'s Marconi-supplied wireless as the ship broke apart and sank beneath the pitch-black, super-cold 28-degree salt water of the North Atlantic that awful April 1912 night. Bride survived; sadly, Philips did not.

Meanwhile, Harold Cottam, the twenty-one-year-old Marconi radioman on the mail ship *Carpathia* received the *Titanic*'s CQD (distress) midnight message and immediately alerted his captain, Arthur Rostron, who instantly ordered a team of stokers to build a full head of steam. By 12:20 a.m., the little mail ship was bravely racing to the rescue at its full, 17-knot speed. Dodging icebergs in pitch darkness for 107 kilometers, it got there at 3:30 a.m., just an hour after the *Titanic* went down.

By then, hypothermia had killed over 1,500 of those poor souls in the water. Most expired in less than an hour. Yet, miraculously, 706 living were pulled into lifeboats by the heroic *Carpathia* crew. Radio had helped save them!

This rare photo (circa 1911) shows the radio room of the RMS Titanic, with its Marconi equipment. The operator was not identified, but he was trained and employed by the Marconi Company (not White Star Line).

So, the *Titanic*'s awful death toll was reduced by one-third because radio made life-saving possible. Much notoriety was piled onto Marconi as the result of overblown stories in the press. His celebrity skyrocketed. Only a few scant lines said anything about Tesla. Yet, it was the latter's radio invention that clearly deserved a heroic place in history.

A quarter-century later, Marconi died, in 1937. Afterward, four of his U.S. patents were immediately invalidated and revoked by the U.S. Supreme Court, because they had been anticipated by Nikola Tesla's years-earlier patents. New patents were issued in Tesla's name, but only after he had passed away, penniless and alone, in 1943.

In fairness, it should always be remembered that Nikola Tesla actually invented radio. His was the true inventive genius. Marconi brought a different kind of "genius" to radio—for implementing it, marketing it and creating a working service for the public.

So, in an odd twist of fate, both these men seem to have been needed for a higher purpose—to save many lives in the *Titanic* disaster. They both deserve to be remembered—and cheered—forever.

Tesla's Radio Remote-Controlled Boat

I must mention here that Tesla also developed inventions that extended radio to actively control devices. To demonstrate this capability, he created a radio remote-controlled model boat, which he patented (U.S. Patent 613809A) and exhibited at the Electrical Exhibition in 1898 at New York's original Madison Square Garden.

The boat was controlled by radio waves generated by a spark-gap transmitter hidden in a control panel near where Tesla was standing. The impulse radio waves it generated were picked up by an antenna on the model boat, which was floating in a large water-filled tank. Within the boat, the electrical pulses were conducted through an early form of radio detector, called a "coherer," invented by Edouard Branly in 1890. This was a glass tube filled with iron and brass filings.

When a radio wave impulse passed through the coherer, the filings stuck together, and current from a battery was able to flow to steering or acceleration motors. These activated to control the boat. The mechanism also struck the glass tube, shaking the filings and causing them to "de-cohere," which stopped battery current flow, until the next radio burst.

Most who saw the demonstration could not begin to fathom "radio." A few even accused Tesla of hiding a small trained monkey in the boat as its "operator." But it really was radio remote control, and he really did it in 1898!

Tesla's radio boat (1898)

7

Radio Gains a Voice (1900)

Let us now pay homage to the genius of Reginald A. Fessenden. His mother was Canadian and his father American, so he regarded himself as being both. We owe this brilliant inventor high praise and extraordinary honor, for it was Fessenden who invented voice transmission by radio, starting in 1900. Earlier, Fessenden had taught college-level mathematics and engineering at Purdue and later in Pittsburgh. But he yearned to do electrical engineering, so he moved to New York and worked for Edison and later for Westinghouse.

He began working on wireless equipment and devised spark-gap transmitters and receiving apparatus. His goal was voice transmission, and neither the raspy spark-gap transmitters nor the coherer detectors of the time could detect complex waves like voice sounds. To improve matters he devised the electrolytic detector, which used a cats-whisker, dipped into a liquid, rather like a crystal detector. Its sensitivity was far greater, and it was adopted by the U.S. Navy.

Seeking further improvements, on a cold December 23 of 1900, Fessenden and his assistant, Alfred Thiessen, set up equipment in two shacks with 50-foot towers, one mile apart, on Cobb Island, Maryland. Thiessen went to one shack with his electrolytic detector receiving equipment, while Fessenden and his spark-transmitting equipment went to the other. There was a landline telegraph link between the two.

GEORGE J. WHALEN

Reginald A. Fessenden

After several tries with poor receiving results, the steam-engine generator powering Fessenden's spark-gap transmitter suddenly began to run faster and more smoothly than earlier. He grabbed the mike and said, "One, two, three, four. Is it snowing where you are, Mr. Thiessen? If it is, telegraph back and let me know."

Thiessen heard him clearly. It was the first voice message ever transmitted by radio.

Emboldened by this success, Fessenden set up the National Electric Signaling Company. He built a tower at Brant Rock (near Boston) in Massachusetts and erected another tower on the coast of Scotland.

In January 1906, he tested his path by sending a two-way radio-telegraph message across the Atlantic Ocean. Remember that, in 1901, Marconi had only sent a one-way, incomplete spark-gap message from Newfoundland to England (only the letter "S" was discernible; the rest was obliterated by static).

By contrast, Fessenden now had workable equipment and stations of his own design with full-message capability. His aim was to transmit two-way voice messages across the Atlantic and make Marconi's telegraphy obsolete.

This stunning plan brought more development work that continued over a period of six years, culminating in his actually transmitting the very first amplitude-modulated (AM) radio "broadcast" on Christmas Eve 1906, from Brant Rock.

The content was homemade and via a carbon microphone. Fessenden spoke, did a Bible reading and played the violin, and his family sang carols. This program was actually heard by radio operators aboard a number of U.S. Navy ships and United Fruit ships off the east coast.

This incredible event marked the origin of all AM voice radiotelephone technology—as well as the beginning of broadcasting. What makes this even more remarkable was that it took place in what was still the era of early spark transmitters, but of much higher frequency, and with better receiving detectors.

Fessenden brilliantly knew that to maximize sound clarity and fidelity, he needed as pure an RF carrier, as close to a sine wave, as possible. Years earlier, he had worked in the original Edison General Electric Company and still had many talented old friends and associates there, with added freedom now that Edison was gone.

So, Fessenden contracted with General Electric (GE) to build him a high-frequency, 50 kilohertz (KHz), rotary spark-gap alternator-transmitter. This large and heavy mechanism was developed and fabricated at GE by the gifted engineer Ernst Alexanderson, and was delivered to Fessenden at Brant Rock, in August 1906.

This unique transmitter produced near-continuous sine waves of high frequency. The near–sine wave RF energy it produced was radiated by the 402-foot-tall Brant Rock antenna.

For sound input, Fessenden connected a carbon (resistive) microphone in series with the battery supply, so that the current controlling the spark amplitude was varied in proportion to the voice or other sound striking the mike diaphragm. Thus, radio waves of good purity were produced and amplitude modulated, faithfully reproducing the original sound, and heard in the headphones of ships' radio operators and other listeners. They were stunned to hear voices and music instead of the hiss of sparks!

Fessenden had alerted the Navy and other radio-equipped ships in advance of his broadcast on that Christmas Eve. His transmitter frequency had reached 88 KHz and his broadcast was heard. It was repeated on New Year's Eve.

But, despite the brilliance of Fessenden's invention, Marconi's powerfully influential connections in industry, government and royal quarters ruthlessly blocked him at every turn.

He was denied permissions by the British and Canadian governments to set up commercial stations, because these governments had vested interests in Marconi's established operations. This corruptly choked off financial support, and Fessenden's backers fled—with his patents.

Ernst Alexanderson's GE alternator was an electromechanical transmitter capable of producing 500 watts of continuous sine wave and 50 KHz RF energy when it was delivered to Fessenden's Brant Rock lab in August 1906. By Christmas Eve, its output frequency had been increased to 88 KHz. (U.S. Patent 1,008,577 was granted in 1911.)

By 1928, a weary Fessenden received a final settlement, for his interests, of almost one million dollars from the Radio Trust of America, which then included most of the larger makers of radio transmitters and receivers. He retired to Bermuda and passed away in 1932. Meanwhile, radio broadcasting entered a "golden age" and became a giant, multi-billion-dollar wireless industry ... and wireless had a voice!

8

Vacuum Electron Tubes Arrive (1904)

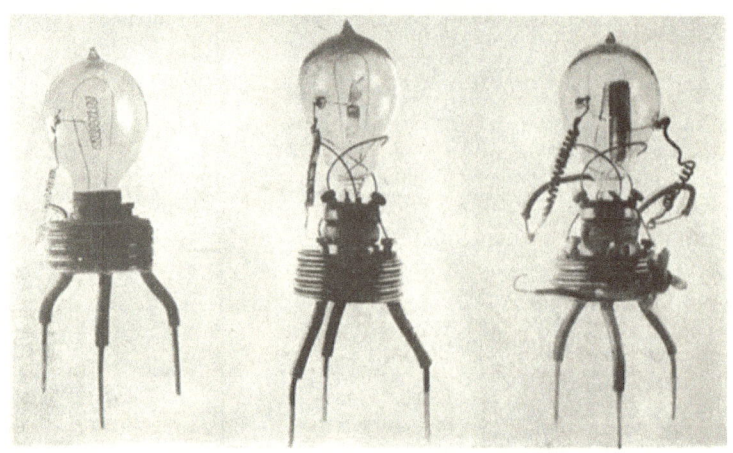

Three versions of John Ambrose Fleming's diode—a two-element thermionic vacuum electron tube

With John A. Fleming's invention of vacuum electron tubes in 1904, the new science of electronics came into existence. Fleming invented the glass-tube, two-element (cathode and anode) diode. It could be used to rectify (change AC to DC), because it allowed current to flow in only one direction. This also made it useful as a detector of modulation in radio receiving devices.

When Lee de Forest added a grid between the cathode and anode, the triode vacuum tube was born, and patented, in 1906. Because the grid allowed a small voltage change to control a corresponding, but larger, current flow between cathode and anode, the amplifier was born. De Forest was never able to explain exactly how his invention worked, which raised doubts about his authenticity as an inventor. Also, although tubes were proven to work best when fully evacuated, de Forest insisted that some gas remain inside his so-called audions. No benefit of this was ever proven.

9

One Inventor Gave Us Regeneration, the Oscillator, the Superheterodyne Receiver, and Frequency Modulation (1912 to 1933)

Edwin H. Armstrong: regeneration (1912), the oscillator (1914), FM transmission (1933), the superheterodyne receiver ... and more!

The invention of the vacuum-tube triode led to the amazing discovery of positive feedback and the oscillator circuit by Edwin H. Armstrong. He was a Fordham University student of Dr. Mihajlo

(Michael) Pupin, a brilliant Serbian-American physicist who later founded the electrical engineering and mathematics departments of Columbia University and was president of the Institute of Radio Engineers.

Young Armstrong became aflame with enthusiasm for radio through Pupin's mentoring. This included learning about "Pupin coils," one of his teacher's many patented inventions. These were used to greatly extend the range of early long-distance telephones. Dr. Pupin also made a great gift of a costly (at the time) triode tube to his student. The young Armstrong used this treasure to experiment with building and trying a variety of circuits and ideas in the attic workshop at his parents' Yonkers, New York, home.

In 1912, he ingeniously used positive feedback, or "regeneration," to greatly increase the amplification of his one-tube circuit from a gain of 20 to perhaps a gain of 1,000. By increasing the positive feedback in the circuit, he found he could cause it go into oscillation. If an adjustable tuned circuit comprising an inductor and a capacitor was also included in the circuit, the frequency of the oscillator could be changed. So, this circuit could be used as a continuous-wave radio transmitter. Armstrong spent the next two years carefully documenting his inventive experiments in regenerative circuits and, in 1914, he received U.S. Patent 1,113,149.

Although it was a well-deserved triumph, four other claimants, including de Forest, challenged the award. While the patent was still being contested, Westinghouse Electric purchased options on Armstrong's patents for $335,000, with an additional $200,000 due to be paid if the regeneration patent was decided in his favor.

It was, but de Forest brought another suit in a Washington, D.C. court, which inexplicably reversed the decision and gave the regeneration patent to de Forest. It was a shocking decision, universally condemned by engineers and enthusiasts everywhere. Armstrong went on to independently develop the superheterodyne receiver, which became the standard design concept for virtually all forms of radio and television receivers throughout the twentieth century. It totally eclipsed regenerative radios. (History now credits Lucien Levy of France with having also developed the design and with patenting it overseas in 1917, the same year as Armstrong.)

Armstrong's oscillator was used in countless radio applications, as well as many vacuum tube generations of radio transmitters. It also inspired oscillator designs invented by Ralph Hartley (1915), Edwin Colpitts (1918), and others. Soon, this marvelous circuit component found its way into most radio and telephone circuits for communication and long distance. Armstrong continued to lead the way in inventing circuits that achieved huge milestones in communication.

One of his proudest, most singular accomplishments was the development of frequency modulation, which not only offered amazing audio frequency fidelity for broadcasting, but also became the de facto standard for interference-free public service and fire and first-responder communications as narrow-band FM (NBFM). (A point: FM was invented by Armstrong in Yonkers, New York, and patented in 1933.) Patent suits and legal pains were a constant depressant to the great man, and it was said that these pressures contributed to his untimely death in 1954.

10

Two-Way Voice Radio Communication Comes to Cars (1920s)

This photo dates from 1924. It shows a very early experiment with two-way mobile radio-telephone equipment (made by Western Electric or by AT&T/Bell Labs).

As early as 1922, experiments with two-way AM voice radio communication were going on in the United States. A few police forces and intrepid amateur radio operators outfitted their early cars with home radio receiving and transmitting equipment. But the bulky equipment and batteries left little room for the driver and none for passengers! A 1924 photo of an experimental radio-telephone car, believed to be then being tested by AT&T, shows a Buick touring car with a radio installation, an operator wearing headphones, and a large loop antenna, which seems sized for lower-radio-frequency work. The mode was believed to have been AM, which was highly susceptible to interference and ignition noise, so this test was conducted at a quiet beach location.

The value of having mobile voice communications was easily recognized. But natural and manmade noise and interference bedeviled "radio cars," so messages were hard to receive. But, as stated earlier, Edwin Armstrong, the young radio enthusiast from Yonkers, had ideas with far-reaching potential. One was that modulating the frequency of a signal could one day be used to make messages, music and other broadcast sounds immune to the buzzes, lightning crashes and squeals that beset AM. Narrow-band FM has become widely used in mobile radio.

The Great Depression (1929 to 1939) slowed or stopped American car production for a time. Then, World War II (1939 to 1945) intervened, and war production took over.

World War II was fought using vacuum-tube communications equipment, and the importance of reliable communication was proven daily. Tubes needed high operating voltages and currents; they were fragile. Something more rugged was surely needed, but solid-state devices were still years away. So, tubes were used. Then, peace came in 1945.

11

Mobile Radio-Telephone Service (1946)

An AT&T installation of a mobile 50W radio-telephone transmitter and receiver fills the trunk space of this 1946 auto.

First in St. Louis, Missouri, then in Chicago, Illinois, AT&T/Bell Telephone started installing heavy, 50-watt Motorola and Western Electric transmitter-receivers in the trunks of subscriber cars and on highway trucks. These were VHF FM mobile radio-telephone systems, operating on the 30–40 MHz public service band. They communicated on two frequencies (half-duplex) up to 25 miles, to large-antenna central base stations connected to the public switched phone system. So, two-way, analog phone calls could be made from the road and received over significant distances. But, because of the high power levels used and the analog nature of the systems, there simply were too few frequency channels to go around (perhaps only twenty-two for a large city).

So, due to interference and delays making calls (because all channels were busy), users were often unhappy. Unfortunately, AT&T expanded this analog mobile radio-telephone service to other Bell System companies. The high RF power used always meant channel-capacity limitations, interference and unhappy users. Callers often had to wait thirty minutes or more for a channel so they could make a call. They also disliked the space-hogging size of the car- and truck-borne equipment.

12

1G Voice-Only Analog Cellular Car Phone Service Arrives (1947)

To counteract the interference and busy-channel problems plaguing their car radio-telephone business, Bell Labs came up with a fresh solution—situate an abundance of antennas and lower-power transmitters and receivers on buildings and towers around town, clustered in small areas (cells); also, use lower-power, limited-range radio transmitters in cars. All fixed cells provided automatic frequency-switching and interconnection to the public switched telephone network. With this cellular plan, a car could drive into a "cell" area so its radioed call would be picked up at the nearby receiver site and put out on the phone line to the called party. The reply would then be transmitted back to the car. As the car quickly drove out of range of a cell site, the signal would automatically switch to the next nearby site, on a different frequency. This would continue as many times as need be, as the car drove through one cell area after another. The low power and limited range reduced interference and allowed channel frequencies to be reused from one cell to another. This effectively increased the number of calls that could be handled at one time. But it still meant you could only use radio-telephone service in the car.

13

That's Right! Cellular Was Invented for Your Car ... Really!

Cellular car radio-telephone service was patented in December 1947 by Douglas Ring and W. Rae Young, both Bell Labs engineers. They recommended tower-mounted antennas connected to base stations. Joel Engel and Philip Porter of Bell Labs also proposed that each cell tower use directional antennas to reduce interference between cells, to increase overall call capacity and for smooth hand-off of calls between cells.

This patented cellular technology had been rolled out by AT&T in several U.S. cities by 1970, but it was strictly for use as an in-the-car radio-telephone service.

Here's why.

The cellular concept had been conceived as a novel way to reuse and raise the number of available in-car radio-telephone channels so that more cars and trucks in a given geographic area could use mobile phones without interference.

There was no idea then that a personal cell phone would ever exist! Besides, everything used tubes, and standard-size parts. So, things stayed that way for quite a while. In fact, Nikola Tesla's predicted personal, portable, battery-powered, mobile radio-phone that you could easily carry in your pocket would come only later, really as the sum of many individual ideas and successes by a number of truly inventive geniuses and engineers. Today, the evolution and convergence of these mobile radio-telephones into smartphone pocket computers and internet data devices has led us to the 5G phone and data network, with unmatched capabilities.

Let's see how that began.

14

A Special Tribute to Al Gross, Inventor of the "Handie-Talkie" (1938)

A soldier demonstrates the use of the SCR-536 "Handie-Talkie." (Motorola photo)

So, where did the personal portable cell phone come from? The story started in 1938, when Irving ("Al") Gross, W8PAL, an American ham and engineer, invented a unique handheld, battery-powered "radio transceiver."

It was some accomplishment for its time. Few parts were miniature back then ... and it used tubes. But it was brilliantly meant to be easily carried by a soldier or downed pilot, and it was able to exchange voice communications with aircraft and other stations up to 30 miles away. Gross worked with the Office of Strategic Services (the predecessor of the CIA) during World War II, and his secret invention was responsible for saving countless lives.

The SCR-536 "Handie-Talkie" shown in the photo above was first produced by Motorola in 1941 and widely used thereafter.

When the war ended, Chester Gould (famed creator of the *Dick Tracy* comic strip) is said to have visited Gross in his lab. He came away with an idea for a two-way wrist radio his comic-strip detective would use to fight crime. It's also been suggested that Gene Roddenberry might have been inspired by that idea to create the small "communicators" used by the *Enterprise* crew in his legendary TV series *Star Trek*.

Al Gross went on from his ingenious handheld radio creation to develop the cordless telephone and pocket paging systems. His efforts inspired the millions of personal cell phones that would come later. For his lifetime achievements, he was honored by Massachusetts Institute of Technology with the prestigious MIT Lemelson Award in 2000. When you next use your cell phone, you'll experience what he alone thought possible, almost a century ago. (Thanks, Al, and 73.)

15

Spread Spectrum: An Amazing Invention by a Remarkable Woman—Hedy Lamarr (1942)

Hedy Lamarr, actress and inventor

Hedy Lamarr was an attractive, famous movie star during the World War II years. She was born Hedwig Eva Maria Kiesler in 1914, in Vienna, Austria, and emigrated to the United States, where she became a naturalized citizen and was signed as an actor by movie mogul Louis B. Mayer. But she was also an inventor. Lamarr learned that radio-controlled torpedoes used by the U.S. Navy had a vulnerability: their receivers could be jammed by an interfering radio signal, sending the torpedo off course.

Loving a challenge, she came up with a brilliant "timed-frequency-shifting system." It changed the guidance signal radio frequency at intervals and also automatically retuned the torpedo receiver to each new frequency. Thus, the torpedo received an uninterrupted signal. This thwarted any attempt to jam the signal, since the jammer could not know each new frequency to which the signal had shifted. Lamarr was assisted in developing her invention by her friend, composer and pianist George Antheil. Together, they built a working model, using a miniature player-piano mechanism to control the radio signal frequency-shifting. They patented it in 1942. But wartime military security kept the invention a secret. Their patent expired and fell into the public domain in 1959. But this ingenious solution endures today. Spread spectrum technology is used in cellular (as CDMA), public service radio, Bluetooth and Wi-Fi systems to ensure privacy of communication. It is a modern form of the brilliant system invented by this remarkable team and was issued U.S. Patent 2,292,387.

16

Bell Labs Scientists Invent the Transistor (1947)

Bell Labs scientists (left to right) John Bardeen, William Shockley and Walter Brattain co-invented the transistor in 1947 and shared the Nobel Prize in Physics in 1956

Just before the mid-twentieth century, revolutionary solid-state semiconductor technology appeared. The transistor (named by Dr. John Pierce, head of the Labs) emerged from Bell Labs, invented by a trio of physicist-geniuses—John Bardeen, Walter Brattain and William Shockley—who worked under his direction. Pierce coined the name from *trans*fer and res*istor*, creating one word that nicely summarized both functions of the device. It quickly found its way into automobile radios and electronic devices, displacing fragile vacuum tubes. The world changed! But the germanium transistor was just the beginning.

Acceptance of germanium transistors in other applications was slow. Ed Cole, the legendary engineering head and, later, president of General Motors, opposed using early germanium transistors in cars, believing them inferior in reliability to electromechanical relays. His opposition helped lead to silicon semiconductors and the development of MOSFETs (metal oxide semiconductor field-effect transistors). These were rugged, reliable and low-cost. And they led the way to "integrated circuits" and large-scale integration.

17

MOSFETs and Integrated Circuits Open a Digital Door to the Future

As mentioned earlier, William Shockley was one of the three co-inventors of the transistor at Bell Labs in New Jersey. Many scientists had joined the Labs staff, working with Drs. Shockley, Bardeen and Brattain, and much commendable work was done in enlarging our knowledge of materials and solid-state physics.

But, in 1956, the ill health of Shockley's mother caused him to leave Bell and move across the United States to Mountain View, California. There, he started Shockley Semiconductor Laboratory in Santa Clara County, where he pursued the development of silicon transistors as an alternative to the use of germanium, the original material. Shockley and his staff believed silicon was a better, more rugged and far less costly choice.

But there were quirks in silicon's surface conductivity that stymied progress. Experiments failed often and tempers flared. Frustration led to ever-more-frequent disagreements between Shockley and eight of his top scientist staffers.

Meanwhile, back at the New Jersey Bell Labs, Dr. Mohammed Atalla was also working with silicon. Experiencing the same kinds of failures that bedeviled Shockley, Atalla and his co-inventor, Dawon Kahng, created a brilliant surface oxide "passivation" process that overcame silicon's problems. They then went on to invent silicon field-effect transistors, an entirely different family of transistor devices from the original germanium point-contact and junction transistors.

These they named metal oxide semiconductor field-effect transistors, or MOSFETs, for short. But there's more.

Inexplicably, Bell Labs chose to fund other work and overlooked the MOSFET, so it languished, wasting time. Shockley, meanwhile, was trying to induce old Bell Labs co-workers to join him in California, but try as he might, none would leave the East Coast.

Shockley's ongoing tantrums and anger had caused him to lose eight brilliant solid-state scientists from his lab. This happened just as Sherman Fairchild decided to open a new semiconductor device subsidiary in California. Fairchild Camera and Instrument Corporation in Syosset, New York, had long been an established provider of technology to commercial and military customers.

What moved him to this new action was attending a conference in 1957, where he heard Robert Noyce make a passionate, visionary presentation about his concept of a silicon-based integrated circuit and his belief in its powerful future. Noyce had seven other colleagues with him—all of whom had been together at Shockley Semiconductor and all of whom had resigned en masse due to friction with Shockley.

Sherman Fairchild already knew of Jack Kilby's concept of a germanium integrated circuit. Its advantages were along the same lines as those of Noyce's silicon integrated circuit, but the implementation differed. Thus, both men had received separate patents, but Noyce is considered a co-inventor of the integrated circuit. Sherman Fairchild was well aware of the genius at his disposal. So, he agreed to create Fairchild Semiconductor with Kilby and Noyce as its leaders.

Fairchild Semiconductor made silicon transistors and produced its first four-transistor integrated circuit in 1960. By 1964, Fairchild had a going business in integrated circuits. A gigantic manufacturing improvement in transistors was developed by Jean Hoerni (one of the eight). This reduced the MOSFET transistor cost, improved performance and gave much higher reliability.

18

Moore's Law, Digitalization, and Technology Growth

Gordon Moore, engineer and businessman (and another of the eight), teamed up with Robert Noyce in 1968 to co-found Intel Corporation. Their hot product was integrated circuit memory chips. In these, silicon transistors were formed directly on the substrate, and Moore recognized that increasingly high-quality "art" could produce smaller and smaller transistors on a specific-size chip, thus increasing its storage capacity. More memory per chip meant more power per computer.

This canny observation in 1965 led to what has been called "Moore's Law." Though it is not a physical law, its impact has been truly profound for over half a century. Moore noticed that the change in the computing power of a memory chip doubled about every two years, because more transistors could be packed onto the same chip substrate. Therefore, if older analog electronic circuits could be "digitalized" and, instead, implemented with the predictably growing power of integrated circuits, the mighty growth predicted by Moore's Law would automatically be conferred as a bonus upon the digital circuit.

This has been a force in expanding the capabilities of integrated circuits and an incredible driver of new technology performance. The modern digital electronic technology used in 5G has been upgraded through many generations by the unseen hand of Moore's Law.

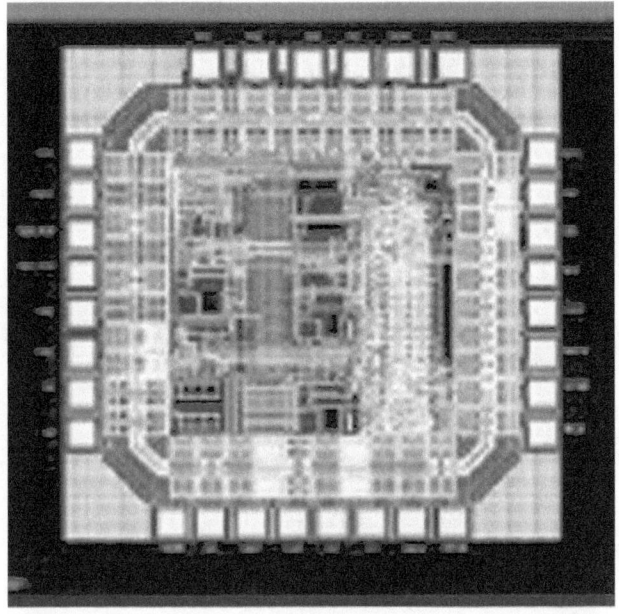

A modern, very-large-scale-integration integrated circuit chip

19

Motorola Invents the First Handheld Analog Cellular Phone (1973)

Dr. Martin Cooper shows how he used Motorola's first handheld analog cellular phone, the DynaTAC 8000, to make history in 1973, when he placed the first handheld "1G" cellular phone call to a rival, from a New York City street corner.

Martin ("Marty") Cooper, a research executive at Motorola, was an ardent fan of *Star Trek*. He idolized the "communicators" that the crew of the *U.S.S. Enterprise* used on the TV show and, one day, challenged his tech staff with the question "Why can't we make a small radio-telephone like that? Everyone could carry one in their pocket or purse and use it everywhere?"

This echo of Tesla's 1926 prediction was quite a challenge for the time. Subminiature components and integrated circuits were still in the future. And getting reliable transmitting and radio receiving up in the frequency range of 900 MHz with reasonably priced devices was still a challenge. But Motorola engineers were up for it and were able to create the first handheld personal analog cell phone.

Dubbed the "DynaTAC 8000," this handheld prototype mobile cell phone was used on a street corner press demonstration on April 3, 1973, in New York City. Surrounded by reporters, Cooper called his rival, Dr. Joel Engel, who headed cellular development at AT&T/Bell Labs, and said, "Hello, Joel. I'm calling you on our brand-new handheld cellular personal telephone from a street corner in New York. I wanted you to be the first to hear it."

The big handheld weighed 2.2 kilograms (over 4 pounds), had a nickel-cadmium (NiCad) battery with a life of twenty minutes and required ten hours to charge. The phone was not made publicly available until 1983, and even then, it was priced at $3,999. It was analog, had no data features and operated on 900 MHz, and its open signal was accessible to anyone.

20

2G Digital Networking and the Digital Cellular Phone Revolution (1980s)

As the 1980s arrived, AT&T analog mobile cellular services continued to be strictly deployed as "car phones." The old-tech analog car phones were still about the only mobile communications service available from AT&T and its Bell subsidiaries, and these analog designs had the same old problems.

Mobile phone services were implemented in a standardized way on all networks throughout the United States so that all phones could work the same everywhere, no matter the state or city. While these cellular car phone services were just about "OK," call capacity was still a big problem. Being analog radio, there was rarely enough frequency spectrum to serve callers, especially in cities and metropolitan areas. Typically, a city might have only twenty-two channels—always in use—so other customers couldn't make their calls. And operating costs were very high. One frustrated Bell company executive said, "The only time we can make any money is when the user is in the car and making a call. At all other times, the analog system is costly and unprofitable."

So, Bell operating companies tried to hand the troublesome analog car cell phone business back to AT&T. But, just then, that corporation was undergoing the eight-year-long legal battle over a court-ordered company breakup that had begun in 1974. So, it was really busy and preoccupied. In 1982, the Regional Bell Operating Companies (RBOCs) were being formed. The car phone problem wasn't first on the AT&T agenda, so it circulated from desk to desk.

Then, in 1988, the problem landed on the desk of director Jesse Russell, a graduate of Tennessee State and Stanford universities. A brilliant young engineer at Bell Labs, he bravely proposed a totally new and exciting solution—go digital! That idea changed the entire character and future of mobile phones. Russell led the first engineering team to devise and introduce digital cellular telephone service. Digital would be the "2nd Generation Network" (2G, for short). And, Russell was right: it was the answer to the old mobile phone problems. Every cellular phone system since then has been digital.

But, there's more: Russell wasn't just the network engineering genius at AT&T; he also proposed taking radio-telephones out of cars—and putting those mobile phones onto people! So, the personal phone and digital cellular network he designed made possible a bandwidth-saving plan using digital voice processing. That increased the number of calls that could be shared on each channel.

Jesse E. Russell, inventor of the digital cellular network, has been called "the father of digital cell phone technology."

This ranks as a truly inventive engineering accomplishment that brought cell phones to the forefront of human communication, solved the old capacity problems and also allowed new digital data services that were not possible with analog technology.

Russell became an IEEE fellow in 1994 for his technology leadership in the development of digital wireless communication, concepts, systems and standards. In 1999, he was also named a fellow of the IEC.

Most importantly, users could now carry with them small, low-power digital pocket cellular phone devices able to transmit and receive voice, emails and text messages. Phones, at last, were out of the car! Small handsets and wider public acceptance made 2G a hit in the 1990s.

21

IBM Pioneers a Personal PDA-Cellular Phone (1991), the 3G Digital Network Arrives (Early 2000s), and Apple's iPhone Debuts (2007)

IBM's "Simon" PDA-cellular phone was the first "smartphone" (1991).

The 1990s brought the era of "personal digital assistant" (PDA) devices that were handy, but not phones. Then, in 1991, IBM patented, and introduced (1992), at COMDEX, something genuinely new. It was the neat integration of a personal digital assistant with a personal digital cellular telephone.

By 1994, IBM commenced marketing its Simon Personal Communicator as the very first "smart" phone (although that name would not actually be coined until 1997). It was a handheld device with a touchscreen interface and it was meant to work with faxes, emails, phones and cellular pagers, as well as other functions. It came with eleven resident programs, which included an appointment scheduler, clock, sketch and handwritten notepad and calculator, a unique IBM-developed stylus input and keyboard with prediction features, facsimile and onboard modem, as well as 1 megabyte of ROM, 1 megabyte of RAM and a 16-bit X86 processor.

The three people most responsible for conceiving and creating the Simon were engineers Frank J. Canova, Jr., Jerry Merckel and the then head of IBM's Florida Lab, Paul Mugge. It was a visionary, pioneering invention. Not only had nothing like it ever existed before; what it was capable of doing vastly exceeded the "technological ecosystem" available in the 1990s. It was simply too advanced for the times.

It was 200 mm x 64 mm x 38 mm, weighed 510 g and was powered by a NiCad battery that yielded eight hours of standby, or constant operation of just one hour. It was assembled by Mitsubishi Electric and distributed by Bellsouth Cellular Corporation, over a network that spanned fifteen of the U.S. states. Some 50,000 Simons were sold, at about $900. The featured touchscreen could send and receive emails, faxes and texts, had full internet access and allowed web browsing, as if you were at a desktop computer. It was amazing. But not everything it needed to succeed yet existed! (Try to remember what the internet was like back then.) So sales stalled. The Simon lasted only until 1994, when IBM had no choice but to pull the plug.

While personal digital phones continued to do well, few offered the brilliant gamut of services and capabilities of IBM's Simon. So the notion of a "smartphone" endured a long thirteen-year hibernation, until 2007, when Steve Jobs introduced the first Apple iPhone.

It incorporated many of the IBM Simon's breathtaking features, but the updated network it would run on by now had matured enough to provide for them.

Basic camera capability was included in the first iPhone, but only low-res, still pictures. Video was delayed. Jobs eased the wait by introducing a new iPhone model, adding new capabilities, every year after 2007. With 3G digital technology, faster connection speeds came in and, ultimately, the capability to stream video and audio as data. These were quite a step up from 2G, but they still required perfecting and occasionally fell short in quality and performance. Three-way calling and call forwarding, as well as short messaging (and price decreases) and increased cellular usage by residential customers, benefited acceptance. The Personal Communications Service (PCS) then operated on frequencies in the 1.8 to 1.9 GHz range.

The iPhone 1 was "locked up" in more ways than one. Apple's distribution agreement with AT&T was exclusive. You could only purchase it from AT&T, and you could only use it on towers of the AT&T network. So, coverage might be OK in some places, spotty in others or non-existent in rural areas.

The Apple iPhone 12 family of handsets has just been introduced. With so many models since the first in 2007, it is getting harder to see truly differentiating *changes* in features and functions. However, the overruling tenet of marketing is that *the product is what the customer thinks it is*. The operating system is Apple's iOS. Things must be going quite well for Apple, as a recent disclosure said that the company has an income of $1 million every 8.8 minutes, or a tad over $163 million per day!

The other class of cellular handsets with multiple manufacturers would be Androids. Based on Linux, the operating system pioneered by inventor Linus Torvald, these handsets have operating clout because the operating system is controlled by Google (or, if you prefer, Alphabet). This company makes $1 million every 17.1 minutes, or $84.2 million per day.

22

4G Digital Cellular Phone/ Data Network Arrives (2013)

In 2013, 4G came along and was a step up from 3G, offering (in theory) a 100-megabyte data speed, good for tablets, smartphones and laptops. So, voice, email, instant messaging, social media and internet were all practical, all with less buffering, better voice quality, improved streaming and faster downloads. It is now widely deployed, after multiple years of build-out and significant expense.

Since it is adequate for most current uses (including GPS), 4G LTE (Long-Term Evolution) is expected to remain in a long supporting role as 5G is deployed. There will be a "5G layer" to handle demanding high-data-volume and high-speed applications, with other, less demanding tasks falling-back to the 4G LTE layer. This triage of data will not be visible to users.

23

An Overview of Wireless Phone and Network Functions

This chapter is a general familiarization tour. The emphasis is on "how it works." Mobile phones are often called "smartphones," "cell phones" or "handsets." Up-to-date units are usually pocket-sized (but getting larger, because they must), light in weight (but getting heavier) and (mostly) made to go with you.

I'll use "smartphone" here as it is most representative of what's available now. (It has true computer power and camera and internet capability, so it really functions as a pocket computer-phone-camera). Regardless of brand, smartphones are typically powered by lithium-ion batteries. These have high capacity, compact size and low weight.

Smartphones use auto-ranging 0.25-watt to 3-watt, low-power transmitters and sensitive receivers. So, these are true mobile wireless (radio) transceivers. They are the modern fulfillment of Nikola Tesla's nearly-100-years-ago prediction of "wireless around the world."

Digital transmission methods are used today. Typically, a carrier gets assigned 832 radio frequencies to use in a city area. This breaks down to 395 voice channels per carrier, plus 42 control channels.

Calls are typically sent via radio and received in duplex (simultaneous talk-and-listen) mode. Duplex occupies two channels on different frequencies (transmit and receive). So, in the simple system example we're using, fifty-six people can be talking at one time. A call is patched through the network equipment of the carrier, into the public switched telephone system. Through that, it is connectable to all other phones and systems everywhere.

Cellular is a marvelous system, using antenna placements, power levels, positioning and digital signal processing to create cell service areas that are relatively small, separate clusters of physical hexagons. The idea is that, as a group of users and their mobile phones move into cells, go through them, then out and into others, multiple conversations can be carried on simultaneously, without interference between them.

This clever plan means cellular frequencies can be reused. This maximizes the network call capacity available to users, so having to wait is very rare. Because the system is "digital" and has high computational capability, smartphones can also handle high-res videos and still photos, video meetings, text messaging, email, internet access, short-range communication (using Bluetooth or infrared) and applications of all types, as well as video games, remote control, and security or baby-monitoring viewing.

The word "smartphone" came into use in 1995. Subsequent generations grew in capability through MOSFET and large-scale integrated circuit technology. Smartphones got smarter due to benefits from Moore's Law (growing on-chip transistor density). Just a single modern smartphone is now said to pack more computer technology power than the entire vehicle that NASA landed on the Moon in 1969.

With digital transmission and coding, the number of available channels goes UP. The three main "standards" used by different carriers to pack calls into the frequencies to which they are assigned are TDMA (time division multiple access), CDMA (code division multiple access) and GSM (global system for mobile communication). Note that a phone made for a given standard will not work on another network, since TDMA, GSM and CDMA differ in caller authentication. Bell Labs TDMA (devised by W. Rae Young in 1947) divides a given, repeating period into tiny "time slices" and shares calls sequentially, "slice by slice." (This is done so quickly that hearers perceive each "share" as continuous speech.) GSM authenticates callers by using customer data digitally stored on the small SIM card in each phone, while CDMA, also from Bell Labs, uses a network-based data verification for each user. All of these standards are in use by different carriers in the United States.

As of 2018, GSM was in use by more than five billion people in over 220 countries worldwide. (The IEEE/RSE awarded Thomas Haug and Philippe Dupuis the 2018 James Clerk Maxwell medal for devising GSM for their contributions to the first digital mobile phone standard.)

Leading smartphone makers (by share of U.S. market) are now Samsung ("Three Stars," in Korean) and Apple. Samsung uses the Android operating system developed out of Linux and managed by Google. The Galaxy line of smartphones is fabricated in South Korea. Apple's iPhone line is fabricated by Foxconn, in Taiwan. Apple uses its own proprietary iOS operating system.

Mobile phones in the United States have long communicated with terrestrial cell towers, placed to give coverage within physical service areas divided into hexagonal cells (as described earlier). Each such cell is assigned to use a different set of frequencies from the adjacent cells, and each is typically served by three towers situated at different points in the cell. The cell towers are hard-wire-connected to each other, to the phone network and to the internet.

Each cell has a maximum number of cellular phones it can serve at once. So, each cell is physically "sized" (in area) depending upon the usage expected to be needed there. Typically, cells in cities will be small in area, because the number of callers is expected to be high. Much lower transmitter powers are thus used, to avoid spilling signals outside the cell and interfering with adjacent cell calls.

As stated earlier, the cellular concept was born in 1947, in the analog era. It was invented (and patented) by Bell scientist W. Rae Young. (He was also a ham, W3KI.) Cellular was originally used to overcome Bell's problems with mobile radio-telephone service in cars and trucks. (The story is told elsewhere in this book.) It also increased call capacity for simultaneous wireless calls.

Say, for example, that a carrier is licensed to use a block of 1,000 frequencies, and these are divided between two adjacent geographic cells. Each cell is assigned a block of frequencies, and every frequency can handle one analog call. Because the transmit power and receive antennas can be optimized to favor only the area of a given cell, its frequencies can actually be reused in other cells that are not nearby.

So, if cell one used frequencies 1–500, then cell two, immediately beside it, could use frequencies 501–1,000. This means that cell three, down the line a ways, next to cell two, can safely reuse cell one's block of 1–500 frequencies without interference, because cell three and cell one are "out of range" of each other.

Of course, when carriers implemented digital networks and put TDMA, CDMA or GSM multiplexing standards into effect, each cell frequency became able to host multiple calls, thus increasing call capacity even more.

From 1983 to 1998, Motorola was the U.S. market leader in mobile phones. Nokia was the market leader in mobile phones from 1998 to 2012. With declining share, Motorola sold its cellular business to Lenovo, which still markets MOTO brand products. In 2012, Samsung surpassed Nokia, selling 93.5 million units versus Nokia's 82.7 million units. Samsung has retained its top U.S. position since then.

The world's largest individual mobile phone operator by number of subscribers is China Mobile, which had over 902 million mobile phone subscribers as of June 2018. Over 50 mobile operators have more than 10 million subscribers each, and over 150 mobile operators had at least 1 million subscribers by the end of 2009. In 2014, there were more than seven billion mobile phone subscribers worldwide, a number that is expected to keep growing as 5G takes hold.

Mobile phones are commonly used to collect location data. While a mobile phone is turned on, its geographical location can be determined easily (whether it is being used or not) using a technique known as "multilateration" to calculate the differences in time for a signal to travel from the mobile phone to each of several cell towers near the owner of the phone.

The movements of a mobile phone user can be tracked by their service provider and, if desired, by law enforcement agencies and governments. Both the SIM card and the handset can be tracked.

24

The 5G Advanced Cellular Phone/ Data Network Starts Up (2020)

Implementation of the 5G network has now begun, but two major restraints remain to be fully overcome as these words are written. First is the tragic global pandemic. The COVID-19 virus injured families, businesses and economies here and worldwide. In the United States, an incredible response by the federal government, first responders, manufacturers and pharmaceutical makers brought all-out efforts to develop tests, therapeutics, vaccines and treatments at historic speed, aimed at mitigating this plague. Three quarters of the year went to shutdowns, business closings, public health and life-saving. But the extraordinary efforts were paying off as the year wound down. A return to normal life was hoped for. Humankind had been under a crushing, preoccupying burden that extended the general availability of 5G into the future. But in a number of cities, 5G was doing demonstrations.

A second restraint is the sheer size and scope of the build-out. The United States is a big part of the task, but the project is planet-wide. It involves the infrastructure of all our population, in all cities, towns and villages. Work also aims to reach most areas of the Earth—some never before touched by telecommunication or the internet. The work has both a terrestrial component and an orbital satellite component. Beyond these are numerous control, computation and artificial intelligence components.

In fact, 5G is the largest, most complex system that humankind has ever attempted. It is an immense job. And it must be completed—and perform—perfectly! Thus, 5G goes far beyond "smartphone communication." It is to be the Big Data Network that will integrate, coordinate and serve all our digital devices, everywhere in the world!

The complexity of doing this is almost indescribable, and it is far beyond the scope of this book to even try. I promised to tell you things no one else has told you about wireless phones and 5G, and that's what I've stuck with. If you want to know *when* 5G will be finished, I simply can't tell you.

What I can tell you is that it is sure to take longer than we have estimated. I will try to explain *why* in a later chapter.

25

You'll Likely Need a New Phone for 5G

The mix of radio frequencies that the various carriers will operate on in 5G will not all be the same. In fact, a given carrier's radio frequencies may range anywhere from 0.6 GHz to 96 GHz. This is because the carrier's "right" to provide service to its subscribers was bought by each carrier at Federal Communications Commission (FCC) auction. The government auctioned off various frequency allocations as they became available. So, different high-bidders wound up getting different frequencies. Since they had also bought other frequencies in earlier-generation phone auctions, each company's allocation is bound to be a different mix. (By the way, if your phone was intended for a 3G network, it is obsolete, for sure.)

Phones for 5G must be able to transmit and receive across the entire spectrum. So, the laws of physics come into play, because 0.6 GHz to 96 GHz is a HUGE chunk of the electromagnetic spectrum. A 96 GHz antenna (in a phone) will be around a millimeter in length. But a 0.6 GHz antenna will be around a centimeter (ten times larger). So, the manufacturer may put as many as five antennas inside the phone case to fit whatever frequency mix it may encounter. This is critical to performance.

Experts say that this network configuration will change how we connect to it as well as how we use it in the Internet of Things (IOT). The IOT includes "smart cities," appliances, heating and cooling systems, industrial equipment and controls, smart manufacturing and enterprise control over data flowing through their own, private networks, as well as supervision and control over autonomous vehicles. The standard is expected to finalize before too long, giving us much to consider.

26

Autos: In Search of Highway Safety

The National Highway Traffic Safety Administration (NHTSA), a branch of the U.S. Department of Transportation, has advocated safety and progressive development of automotive electronic systems that aid the driver since 1966. It has encouraged crash-worthiness in car design and safety measures such as seatbelts and airbags. These changes have measurably helped injury and death statistics from accidents. But, in recent years, as many as 137,00 people have been annual fatalities on our highways.

The NHTSA has been encouraging increasing the number and kinds of driver safety and assistance devices with each new model year. This has led to backup and lane-change cameras, automatic stabilization systems, anti-skid braking, traction control systems, steering and parking assist, and the like.

But, ahead, down the road, is the goal of autonomous vehicles. When 5G is fully implemented, we must have this very-high-speed/low-latency cellular radio communication network fully functioning everywhere.

The thinking behind autonomous vehicles is that by eliminating human drivers from behind auto steering wheels, and using robot-automated, driverless cars, trucks and buses, accidental fatalities and injuries can be greatly reduced. By going over to self-driving vehicles, human shortcomings, errors and variability in judgment can be eliminated from the highway safety equation. The savings in loss of life and injury, experts say, will be immediate. Robots do not drink or smoke pot. They do not concentrate on cell phone conversations while driving. They do not argue with other adults and children in the car. Their attention span is always perfect, and their reaction time is, too. So, the state of the art is moving toward convergence of automotive electronics with the incredible fifth-generation 5G cellular data/telephone communication network now coming into existence.

27

5G Is All about BIG Data!

As explained earlier, though the phone carriers may love to use the term "wireless," which has a certain charm, what they have been promoting is ever-advancing, sophisticated radio-telephone service.

Why? Certainly there's no great need for more cellular voice phone circuits to call grandma or the person minding your dog or to get spam sales calls. No. What is really driving this expansion is our overwhelming, unbelievable appetite for communicating data. Whether it's streaming news, TV, texts, instant messages, photos, stock market data, business operations, shopping carts, games, websites or movies, the sheer volume of digital data now trying to squeeze through existing 4G cellular networks is very close to exceeding their peak capacity.

Every two days, it's said, present networks carry one exabyte of digital data. That's 1×10^{18} bytes (approximately one quintillion bytes—or, if you prefer, one billion gigabytes!)

There is not very much big enough to compare to an exabyte. It is a lot! It has been said that five exabytes would be about equal to all of the words ever spoken by humankind! But you get the picture.

Today, the average U.S. household is said to consume 190 gigabytes (GB) of data per month and has an average of 14.7 cell phones and internet-connected devices. These include tablets, laptops, PCs, game consoles, smart TVs and appliances.

This figure is only going to rise, as the Internet of Things (literally, every kind of appliance and device) is connected and communicates online for information, control and maintenance.

As a rough gauge of "data," consider that streaming an HD feature movie takes around 4 GB. It may sound like a lot now, but experts estimate our household use of data will grow by 38 times in the next ten years!

When the carriers got together in the Third Generation Partnership Project (3GPP) and developed the initial standards for 5G New Radio (NR), the standards said that up to 1 GB per second speed is to be available, with less than one millisecond delay (low latency), with maximum download speed, under best conditions, of 10 to 20 GB per second.

So, if a high-speed, high-volume, low-latency network like that were installed everywhere on Earth, wouldn't you grab on to it for data communication with autonomous vehicles?

Well, that's what is starting to happen.

Carrier networks now moving to 5G are in many cases using it as a supplement to their existing 4G LTE systems, which have taken years to install. (Oh, yes ... full 5G is still "coming," but it will likely take a good while to get here.) Since 5G mobile networks are still in various stages of deployment, the potential for functional 5G-connected autonomous vehicles is still some time away. As carriers are investing billions into 5G networks, research into autonomous driving is accelerating.

According to some sources, extensive investments are needed in the development of machine learning models for interpreting traffic and of infrastructure for low-latency wireless networks. These are just two more of the things we need to roll out 5G fully.

28

5G New Radio (NR) Technical Summary

The group responsible for developing 5G technical standards is the Third Generation Partnership Project or 3GPP. These are all the carriers. Their task in coming up with agreed-upon standards that can work across all of their domains and geographies is truly formidable.

Every carrier is sure to have a brochure or pamphlet describing the particular system configuration that they will use in your locality, and you should study it carefully. There may be substantial differences based on the frequencies they use. The major divisions of frequencies are 450 MHz to 6 GHz, 24.25 to 52.6 GHz, 59.25 to 71.50 GHZ, and 64 to 86 GHz ("millimeter waves").

In the United States, frequencies allocated by the FCC to carriers also depend on those they already own. For 5G, carriers have had to seek out frequency spectrum they now own where they still have bandwidth that can accommodate their 5G work. If there's not enough, they must acquire new, supplemental frequencies in the 24 GHz, 39 GHz or higher-frequency bands. This spectrum is to be cleared by the FCC and auctioned off.

This may leave some carriers operating on frequencies below 6 GHz, as much cellular does now, as well as on higher frequencies. The challenge is that all carriers must be able to work at 5G standards everywhere. So, the diversity of spectrum is bound to impact the size, weight and cost of handsets, but it will also impact equipment design in the infrastructure.

The FCC is repurposing and auctioning off 285 MHz of the invaluable "C-Band" (4 to 8 GHz) for 5G. Nearly all communication satellites presently use 3.7 to 4.2 GHz for downlinks and the spectrum from 5.925 to 6.425 GHz for uplinks. C-Band wavelengths are measured in centimeters, so they're longer—ranging from 7.5 to 3.75 cm. They travel further and penetrate better into buildings. The millimeter wavelengths of higher frequencies don't propagate as far, and they don't penetrate buildings or foliage well. This will necessitate having an abundance of terrestrial fixed-antenna and receive-transmit sites.

29

Satellites, Network Speed and Latency

The original plan approved by the FCC was that, by 2025, carrier spectrum availability changes would be settled. However, that was before the pandemic hit us. So we've probably lost a year. Legal wrangling and other delays are always possibilities as well. One can only wait and hope.

The best advice is to stay tuned for news and announcements. Don't be in too great a hurry to buy your new smartphone. Your old 4G phone will work just fine until 5G availability can be confirmed and trusted. By the way, a new smartphone could set you back $1,399 or more.

The newest 5G spectrum band frequencies are above 24 GHz and 39 GHz. Ultimately, frequencies may even range up to 96 GHz. Greater bandwidth is available up there, in the millimeter wave bands, to accommodate the data speeds foreseen to be needed. But wavelengths that short are about the size of a raindrop—and are said (by some) to be about as fragile. The power to be used up there is going to have to be higher to compensate for absorption and path losses.

Speed and low latency of the network are of prime importance. How quickly data gets from the source to the user is critical. Consider that in the heyday of the 4G network, the peak speed was (yawn!) 10 megabytes per second and latency ran 30 milliseconds! But in a 5G network, the design peak speed is a snappy 10 to 20 gigabytes per second and one millisecond latency!

The need for speed becomes clear when you think of the network providing data to moving robotic cars. Consider a car driving at 60 miles per hour. It is travelling 88 feet per second. So, in a tenth of a second (100 milliseconds), it will go 8.8 feet (106 inches). And, in a hundredth of a second, it will go 10.6 inches. So, latency—the speed at which transmitted data gets to a car, a surgical or industrial robot, or other device—really determines the degree of safety, precision or response that is possible by the system. One millisecond latency is about the figure desired for 5G.

30

Low-Earth-Orbit (LEO) Satellites

While frequencies and other specs are being finalized, the launching of low-Earth-orbit (LEO) satellites also needs to be completed. A great many LEO satellites will be needed to provide good, global signal coverage of all places on Earth. SpaceX and Amazon have been involved in this endeavor. The aim is to provide 5G network accessibility and establish a new, faster, low-latency internet. (Of course, the original (but slower) internet will remain in service, with its 740,000 miles of undersea fiber-optic cable that interconnects the nations of the world.)

SpaceX
SpaceX has Starlink, its proposed constellation of LEO satellites, designed to orbit at relatively low (300–600 miles) altitudes above the Earth and to beam fast network coverage to the surface below. SpaceX initially secured licensing from the FCC to launch 12,000 satellites into orbit. It also has approval from the International Telecommunication Union (ITU) for use of radio frequencies to communicate with 30,000 more satellites. So, SpaceX plans to launch a total of 42,000 satellites into orbit. A number have already been placed for testing purposes. In the fall of 2020, an unspecified number of satellites were removed and replaced by SpaceX. No reasons were immediately available.

Amazon
Project Kuiper is an initiative from Amazon to launch a constellation of LEO satellites that will provide low-latency, high-speed broadband

connectivity to unserved and underserved communities around the world. The company says, "This is a long-term project that envisions serving tens of millions of people who lack basic access to broadband internet. We look forward to partnering on this initiative with companies that share this common vision." The Amazon constellation will use three LEO layers of satellites: 784 in a 590-kilometer orbit, 1,156 in a 630-kilometer orbit and 1,296 in a 610-kilometer orbit.

31

How Long Until We Have "Autonomous Vehicles"?

This is a really BIG question. We are in the process of "growing into" robotic cars. The Society of Automotive Engineers has helped matters greatly by defining the steps we'll pass through on our way to that point.

The following is a quotation from the SAE International Recommended Practice, official document J3016_1—@201806, reproduced here with the organization's consent:

Taxonomy and Definitions for Terms Related to Driving Automation Systems for On-Road Motor Vehicles

This SAE Recommended Practice describes motor vehicle driving automation systems that perform part or all of the dynamic driving task (DDT) on a sustained basis. It provides a taxonomy with detailed definitions for six levels of driving automation, ranging from no driving automation (level 0) to full driving automation (level 5), in the context of motor vehicles (hereafter also referred to as "vehicle" or "vehicles") and their operation on roadways. These level definitions, along with additional supporting terms and definitions provided herein, can be used to describe the full range of driving automation features equipped on motor vehicles in a functionally consistent and coherent manner. "On-road" refers to publicly accessible roadways (including parking areas and private campuses that permit public access) that collectively serve users of vehicles of all classes and driving automation levels (including no driving automation), as well as motorcyclists, pedal cyclists, and pedestrians.

The levels apply to the driving automation feature(s) that are engaged in any given instance of on-road operation of an equipped vehicle. As such, although a given vehicle may be equipped with a driving automation system that is capable of delivering multiple driving automation features that perform at different levels, the level of driving automation exhibited in any given instance is determined by the feature(s) that are engaged.

This document also refers to three primary actors in driving: the (human) user, the driving automation system, and other vehicle systems and components. These other vehicle systems and components (or the vehicle in general terms) do not include the driving automation system in this model, even though as a practical matter a driving automation system may actually share hardware and software components with other vehicle systems, such as a processing module(s) or operating code.

The levels of driving automation are defined by reference to the specific role played by each of the three primary actors in performance of the DDT and/or DDT fallback. "Role" in this context refers to the expected role of a given primary actor, based on the design of the driving automation system in question and not necessarily to the actual performance of a given primary actor. For example, a driver who fails to monitor the roadway during engagement of a level 1 adaptive cruise control (ACC) system still has the role of driver, even while s/he is neglecting it.

Active safety systems, such as electronic stability control and automated emergency braking, and certain types of driver assistance systems, such as lane keeping assistance, are excluded from the scope of this driving automation taxonomy because they do not perform part or all of the DDT on a sustained basis and, rather, merely provide momentary intervention during potentially hazardous situations. Due to the momentary nature of the actions of active safety systems, their intervention does not change or eliminate the role of the driver in performing part or all of the DDT, and thus are not considered to be driving automation.

It should, however, be noted that crash avoidance features, including intervention-type active safety systems, may be included in vehicles equipped with driving automation systems at any level. For Automated Driving System (ADS) features (i.e., levels 3-5) that perform the complete DDT, crash avoidance capability is part of ADS functionality.

Now, here is what the National Highway Traffic Safety Administration (NHTSA) says these six levels mean:

- **Level 0:** The human driver does all the driving.
- **Level 1**: An advanced driver assistance system (ADAS) on the vehicle can sometimes assist the human driver with either steering or braking/accelerating, but not both simultaneously.
- **Level 2**: An ADAS on the vehicle can itself actually control both steering and braking/accelerating simultaneously under some circumstances. The human driver must continue to pay full attention ("monitor the driving environment") at all times and perform the rest of the driving task.
- **Level 3**: An automated driving system (ADS) on the vehicle can itself perform all aspects of the driving task under some circumstances. In those circumstances, the human driver must be ready to take back control at any time when the ADS requests the human driver to do so. In all other circumstances, the human driver performs the driving task.
- **Level 4**: An ADS on the vehicle can itself perform all driving tasks and monitor the driving environment—essentially, do

all the driving—in certain circumstances. The human need not pay attention in those circumstances.
- **Level 5**: An ADS on the vehicle can do all the driving in all circumstances. The human occupants are just passengers and need never be involved in driving.

32

Robocar Design Is Still Evolving

What the previous chapter shows is that we still have a number of automobile generations to go through before we achieve Level 5. In the year 2020, new cars were approximately at Level 1 or Level 2, depending upon how much you paid for them. The key words were "driver assistance packages." These are technologies intended to make motor vehicle travel safer by partially automating, improving, or adapting some of the tasks involved in operating a vehicle. These are described below. Carmakers have come up with a variety of assist packages and prices. But there are differences, so, even if the names of two carmakers' packages seem the same, their functionality may be different.

Driver Assistance Package: Lane Keeping Assist uses a video camera to monitor the road and detect when the vehicle is about to leave its lane, so that it can give the driver a warning. A step up is Active Lane Keeping Assist, which is able to operate the brakes to help keep the car in its lane.

Co-Pilot360 Advanced Suite of Driver Assistance Technologies: This includes automatic emergency braking (with pedestrian detection), a blind-spot information system, a lane-keeping system, rear backup camera, and automatic high-beam lighting control.

Adaptive Cruise Control: This is equipped with stop-and-go and can function at very low speeds, including coming to a complete stop and starting again, making it perfect for preventing rear-end collisions in stop-and-go traffic.

Forward Collision Warning and **Automatic Emergency Braking:** These enhance safety at highway speeds.

33

We've Got a Way to Go Before Autonomous Car Technology Converges with Still-Building 5G Network Technology

We still have three (or so) "Levels" of cars to finish designing, building and testing. And 5G has lots and lots of other "moving parts"—including all of the immense technologies we need to still get working and to bring together. We have to install a gazillion new ground-based antenna and repeater sites (indoors and outdoors, about every 500 feet) for the new global 5G system of cellular high-speed/high-data-volume/low-latency networks. And, while we're at it, we have to build, launch and get working some 42,000 low-Earth-orbit communications satellites. Oh, and we still have to recover from the delays induced in everything by the global pandemic of 2020–2021.

Then there's the teensy problem of getting all those satellites to work as a well-coordinated system, everywhere on Earth! And how will we handle "command and control" with ground stations we haven't yet designed or built? And did I mention all the artificial intelligence we'll need? Wow!

And where's the control software? And the quantum computing systems we'll need? Oh yeah!

And, then, of course, where are the advanced, tested and proven robotics we'll need to control trucks, buses and cars? And, oh, I almost forgot, all the design, infrastructure building, traffic modeling, controls and other stuff we'll need to create "smart cities." And, gee, we also have to fit in an Internet of Things to run all our machines ... and Alexas!

Whoops! I DID forget that we have to put all this NEW stuff on top of an aged (crumbling) infrastructure of roads, streets, highways and buildings that were never meant to handle modern technology and to support whiz-bang stuff!

I don't think that even the best robotic car is going to do too well on a New York street that's full of potholes, with hundred-year-old water and gas pipes below. Guess we'll have to fix all that too!

34

No Good Estimate of a Finish Date ... Yet

To put it as gently as I can, the build-out to the FULL 5G era seems fated to take quite a bit more time than the prognosticators said. The NHTSA figured it at 2030+, but that "+" could be fifteen to twenty years, or longer. We simply can't know now how long it will take to make, perfect and launch such a very, very large-scale technology system solution with countless antenna sites (in the millions) on ground structures, tens of thousands of satellites in low Earth orbit, and everything else needed to create a network able to reliably control millions of cars, trucks and buses and to safely deliver people and goods to destinations everywhere on Earth.

Oh yes. And to give us a new internet able to constantly handle multiple exabytes of data (1×10^{18} bytes). Also a new IOT that will control devices of all kinds in every area of life. Oh, and all this at very high speed and with low latency (about 1 millisecond) so it can be safely used for critical, remote operations (such as robotic surgery).

If this is not enough, you may want to consider that the sheer volume of data to be processed every instant, 24 x 7, will vastly exceed the capabilities of our present highest-speed, highest-volume digital computers. Fortunately, Honeywell, IBM, Rigetti Computing and Google have made astonishing progress in quantum computers, and those will, quite possibly, be liberally needed in the 5G data environment. But they, too, are a bit downstream in quantity. And so is their software. And, of course, artificial intelligence will be widely needed.

Streaming video will be everywhere so we can be assisted, entertained and educated as our technology works.

Oh, and do consider that the incredible data volume to be handled by the 5G network will be implemented on microwave radio frequencies ranging from 6 GHz to 95 GHz. Those higher-end millimeter waves still are relatively new to us as a radio frequency territory, and much remains to be learned about them, in practice, to ensure dependable operation under all weather and operating conditions.

Now, given all this work and variability, think about the overall 5G era for just a moment. What we are seeking to do is build an amazing *convergence* of many *developing technologies,* to create a world in which autonomous robot vehicles, vastly lower numbers of highway accidents and fatalities, unprecedented communication, data and control, and a quality of life that can only be imagined now will become an everyday reality!

A truly gargantuan task, to be sure.

Yet, there are signs we will able to do it. For example, every major U.S. car manufacturer has declared its intention of manufacturing fleets of "robot taxis" and deploying them throughout the country. Want to go somewhere? Say so and, instantly, a robot taxi will come to your door, identify you, whisk you to your destination, charge your account and go away after you alight. Imagine. No parking headaches. No car insurance. No gas bills. No repairs. And, hopefully, no accidents! People will simply be "out of the driving loop," so the chances of injury will be greatly reduced. So says the theory. Also, cars will not be owned and used like today. If you buy an autonomous car, you'll likely send it out daily as a robot taxi so it will earn for you.

Commitment to this scenario seems clear. A General Motors chief executive recently said, "We have all of the necessary assets under one roof," so GM looks to be a frontrunner in the coming brave new autonomous car world. Other car firms are also now announcing their involvement in developing automotive electronic systems that have as their key objective the elimination of human drivers from behind steering wheels, through the use of automated cars and the incredible (once fully deployed) 5G network.

The belief is that robot autos, augmented by other technologies such as artificial intelligence, quantum computers and high-speed/low-latency data communication will eliminate human shortcomings, errors and variability in judgment, thus raising highway safety to new heights never possible with human drivers and bringing savings in terms of lives preserved and injuries forgone. That, at least, is the theory.

After all, robots do not drink, eat or smoke pot. They do not use cell phones while driving. They don't argue with other adults and children in the car. Their attention span should theoretically be perfect, and their reaction time should be, too. Plus, the high speed and low latency of the finished 5G network and its high-tech components should ensure virtually instantaneous communication. Artificial intelligence should always make ideal decisions, and nothing should possibly be able to go wrong. Right?

Again ... that's the theory.

But 5G is still unfolding. AI is, too. And the quantum computer is still on its way to a magnificent future. The satellite clouds and ground-based units of the information network remain to be completed, the backhaul fibers have to be put in, and the new internet and the IOT must be done. Several tons of software need to be written, tested and proven. In fact, all the pieces remain to be put together before they can give us proof that they'll work in prolonged, exhaustive testing. Oh, and by the way, we still have to complete safe, reliable vehicle designs that can comfortably and capably transport and protect human passengers wherever and whenever they wish to travel, with absolute assurance they will get there.

As said (before the pandemic), the industry's best guess at when the 5G robotic car scenario would play out was some time in 2030+, or later. So, we have a possible "then" and "right now." In between is a period in which the expansion of 5G and electronic technologies must bring massive changes. But, clearly, it will all take a while!

Self-serving commercials are popping up on TV from each of the carriers who've managed to get some of the 5G network set up, for calls in one or more of the U.S. demonstration cities. But, just as one swallow does not make a summer, being able to make a 5G phone call does not mean that the entire 5G network is up, available and fully operational ... yet! There is a very big difference between saying "5G is here, now" and "5G is fully implemented and ready."

Go back and look at 4G's history. It took ten years (and more) to get the 4G network adequately implemented, even in populous areas. Critics say that in a great many areas of rural America, 4G still does not exist. They say there are still gaping holes in 4G's coverage map of the United States. Anyone whose smartphone has suddenly gone quiet in a rural area knows exactly what that means. In an instant, you are on your own, cut off from everyone and everything. It's not a pleasant feeling.

Uniformity of reliable 5G signal coverage is a major challenge facing carriers. The gaps, holes and null service areas we have in rural and suburban America also exist elsewhere throughout the world. In fact, 5G poses a tougher challenge, because so much of the bandwidth used may have to be in the millimeter-wavelength frequencies that can extend to 24 GHz, 39 GHz and even 96 GHz. Radio frequency energy at these tiny wavelength extremes does not travel far, does not penetrate buildings, structures or trees, and can be affected by rain, sleet or snow. It is easily reflected and can cause unexpected standing waves and nulls. It is generally regarded as a bit "tricky" to work with. Yet, those high frequencies offer the precious bandwidth needed to carry increasing volumes of big data.

We will also have antennas everywhere. Just about every light pole, signpost, building (outside and inside) and tower will likely sprout 5G antennas and relays. (The rule of thumb seems to be every 500 feet.) Fortunately, antennas (and units) will be smaller than earlier generations, thanks to the shorter wavelengths in use.

Will we deal well with low-Earth-orbit satellites? SpaceX, as well as other firms—OneWeb, Boeing, Spire Global and Blue Origin—have proposed to put up "global constellations" of 5G LEO satellites around the Earth. The present forecast is 50,488, but the number changes often. Launched and placed in orbit at an altitude of 300–600 miles above us, it is said that such a near-space cloud of radio relay satellites could perform essentially the same function as the countless 5G terrestrial antennas and relay sites planned to be installed on buildings, light poles, signposts and other structures in more populous areas.

SpaceX received FCC approval to launch and test LEO satellites, approximately sixty at a time, in flight after flight of its reusable civilian Falcon rockets. To its credit, SpaceX's planning, execution and recovery efforts have been good. The company has now estimated that 42,000 (up from an original 12,000) satellites will be the minimum required to provide internet service and complement the world's 5G needs.

How many launches will that take? SpaceX has said that each relay satellite unit it puts up will weigh about 500 pounds. So, if a full Falcon rocket payload is 60 units, that's a respectable payload of 30,000 pounds per launch. Now, if you break that down to how many launch flights are needed to, you get 700 launches.

That is quite a schedule. It requires lots of manpower. Many Falcon rockets, even if reusable. Manufacturing thousands and thousands of relay satellites. A gazillion dollars. A l-o-n-g time. If seventy launches can be made in a year—a very, very respectable number—SpaceX could grow the constellation by 4,200 relay satellites every year. But, it will still take ten years to complete that part of the 5G network. And that's if everything goes perfectly

And what about ground terminals? Well, yes. People who want to access the SpaceX satellite constellation for internet or 5G use will need a (dish) antenna and a terminal of some kind. This will have a monthly cost (to the user), as yet not specified by SpaceX. (However, one person familiar with the matter has estimated that many people are now paying about $80 a month for satellite TV from faraway, high-latency geostationary sats whose signals are also affected by weather. Until further 5G information is published, this is the best estimate figure available. Meanwhile, Amazon Web Services (AWS) has said it will invest in "a dozen" ground stations (more or less) to upload data to the LEO constellation. No further information on this 5G element is yet available.)

Latency. What About LEO satellite latency? Under ideal conditions, the physics of present geostationary satellite communications accounts for round-trip latency of about 550 milliseconds. (That is over half a second—an eternity by tech standards!) SpaceX estimates the round-trip latency of a near-space LEO satellite will be substantially less. If the ground antenna placement of every 500 feet can be relied upon, latency figures in the single-digit millisecond range should be possible.

Robocars. What about autonomous vehicles? Lots of testing continues. Many successes have been rolled out. Unfortunately, a variety of fatalities, injuries, failures and mishaps have happened, necessitating changes. These problems have so far only been experienced in tests of self-contained autonomous vehicles. Though their safety record is improving, much continues to be done to ready them for network use.

So, what's the rush? Until every part of 5G is settled, our advances into the brave new world some envision may simply have to slow down. This means cars are going to be pretty much as they have been for a while longer. In fact, even after "the big 5G change" comes, there are still going to be lots of legacy ICE cars around. Their systems will work pretty much as they did in the twentieth century. Oh, those cars will eventually disappear from the road ... but it might take years and years!

Autonomous vehicle technology is destined to grow and change hugely as very great artificial intelligence advances and 5G new radio telecom and data network deployment take place worldwide. It will undoubtedly take time and occur in stages. Meanwhile, more and more "safety electronics" will be added to cars.

35

Estimated Car Electronics Costs As Time Goes By

According to industry data provider Statista, the percentage cost of electronics in cars will grow hugely. The percentage cost of electronics in a 1970 car price was about 1.5 percent. But, by 2050, 47 percent or more of a car's price may be for electronics! This year, a typical car's list price is $36,000, and its electronics are said to already account for nearly an informally estimated 36 percent ($12,960) of that price. (These figures are clearly guesswork.)

Driver aid and assistance package costs are largely for more and more electronics. At some point, a thoughtful solution must come to this price spiral, or private ownership may simply yield to robotic taxi vehicles, which have been described earlier. This is another large and intensive matter that will surely require considerable time to resolve.

So, as you can see, the things no one has yet told you about may simply not have been fully figured out ... yet. Each of these things will take time, much thought, inventiveness and practical trials. There is no doubt that it will happen. The only uncertainty that remains is *when*.

Since this means your present car may continue to give you satisfactory service for longer than you'd thought, give yourself a treat and be sure to look at my upcoming book on ICE car automotive electronics. Read about the principles of the time-proven systems of your internal combustion engine car, how they work and how to get the most from them.

Afterword: Is There a Dark Side to 5G?

Is 5G Able to "Spy" on Us?

If you stand back and squint your eyes a bit at the image of 5G, you will inevitably see that—as wonderful as this technology convergence seems—there are some "shadows" in it. These could become negatives to U.S. society.

First is the fully deployed 5G system's capacity to "spy" on you, track you 24 x 7, and PERMANENTLY RECORD every detail of your life. It has been dubbed by some as "a surveillance system," and its very pervasiveness and thoroughness of data-gathering about you and your daily life add up to a total that could be frightening.

The concept of "privacy" will be nonexistent. The strangest part will be that what you think of the system will depend upon your personal attitude toward it. On the one hand, you will be surrounded by "services" and "things being done for you," which will make living near effortless. We will surely have far more cameras, artificial intelligence, sensors, voice command, health, computer power, robotic muscle, software control, recognition, and control over payments and record-keeping than ever before.

The automated Amazon store is a good example. You choose what you want as you pass through the store. The system tracks your selections and your satisfaction and it tots up your finances, where and how well you live, where you travel, where you stay, who you are with, and so on.

Will you accept such real and potential privacy incursions, in exchange for faster service in automated stores, quicker downloads of games and movies, or the convenience of driverless autonomous vehicles? It's a quandary to folks who value their privacy, liberty and the freedoms guaranteed by our Constitution.

When "the network" always knows where you are ... what you are doing ... who's with you ... where you are going ... and, maybe, why you are going there, it then is but a step to a robot coming up and saying, "Your papers, please!"

Does that sound like totalitarianism? It is! One need only follow (to the extent possible) the way 5G has been applied in communist China and Hong Kong. There, the 5G system is much further along in deployment than in the United States. But such messages as are leaking out inform us of the communist government's total authoritarian control of the people, repression of protesters and ever-tightening control over communication. Our experience with China in the Hong Kong riots and COVID-19 pandemic also provides illuminating insight into what could happen. Keep abreast of developments in our system and be skeptical of anything that might decrease your rights. Security is comforting, so long as you are not held against your will.

Is 5G Free of Health Risks?

Then there is the long-unsettled question of 5G health risks. Questions have lingered because no one in science or medicine has yet *proven* that 5G electromagnetic radiation is conclusively safe. Instead, technical and medical sources like the FCC cling to tired, vague rhetoric that "RF energy radiated by radio devices in the new network are non-ionizing radiation, and so, are generally regarded as safe." This rather "lawyerly" statement is a coin of opinion that constantly flips back and forth. The root cause is said to be people's bad reaction to the word "radiation." It conjures frightening mental images of wartime destruction of whole cities by atomic bombs, as well as the "nuclear winter" devastation of Chernobyl, caused by its deadly reactor meltdown.

Actually, there are two kinds of radiation—ionizing and non-ionizing. Of these, ionizing radiation is outright hazardous. It's associated with things such as x-rays, atom bombs and uranium. All release high-energy waves and particles that can penetrate the human body, strip atoms off DNA and lead to cell mutations and malignancies.

Then, there's non-ionizing radiation. It is the lower-energy radio frequency kind that is used in radio, TV, ham radio, CB ... and cellular phone communication. Its lower energy levels are supposed to mean its radiation waves won't damage cells—though it can produce some tissue heating.

There have been some studies of RF radiation (on rats and mice). The results have been universally bland, usually saying non-ionizing radiation is "generally thought safe." (Where have we heard that before? Just yesterday, doctors smoked, and the tobacco industry loudly claimed tobacco did not cause cancer. How did that end?)

While the short answer is that, until now, non-ionizing radiation has been thought of as essentially harmless, the longer answer is—we simply don't know for sure. Not enough real scientific testing has been conducted to conclusively prove that 5G RF radiation is safe.

In May 2011, the World Health Organization stated that mobile phone use may possibly represent a long-term health risk, classifying mobile phone radiation as "possibly carcinogenic to humans," after a team of scientists reviewed studies on mobile phone safety. The mobile phone is in category 2B, which ranks it alongside coffee and other possibly carcinogenic substances.

Lastly

In recent months, a peculiar story has come to light. It is that bees, the insects the world relies on to pollinate food crops, seem to be adversely affected by 5G RF. They are said to have trouble finding their way back to hives. What do you make of that? It might be "fake news," but dare we risk it? If we lose them, agriculture ends and the world starves. Shouldn't stories like this be vetted by credentialed scientists? How long could it take? How much could it cost? Once 5G is fully implemented, it will be on 24 x 7 for a very long time. Why not know whether stories like this are true or false before the switch is activated?

When you consider the large amount of radio frequency energy likely to be radiated globally—across the living and workspaces of unborn children, babies, youngsters, women and men of all ages and physical conditions, pets and other animals (and, yes, even the bees that pollinate our food crops), seven days a week, twenty-four hours each day—it seems absolutely reasonable that we should have good, provable medical, engineering and scientific findings that 5G is 100 percent safe.

Epilogue: Here's to Your Future!

Exactly when 5G will be fully implemented in every respect is anybody's guess. The task is beyond huge, and the potential is indescribable. But the effort is worth it, for it will establish a new, upgraded standard of living for all the world's people. All we need do is confirm that it is truly, provably safe.

I hope you have enjoyed this book and that it has done the job I intended—namely, to share with you things nobody ever told you about how we got to where we are on our way to our 5G network future.

Three things helped. First, I was locked down by the virus, so I had plenty of time to research (and I needed it!). Second, I was born in 1939, so I actually lived through some of the incredible "electronic history" I've shared with you. Third, because the right (and best) woman in the whole world married me, I have been encouraged to keep finding and writing stories about real people (like you and me) who just happened to be heroes of the moment in their remarkable actions, discoveries or inventions. Thank goodness for them! I hope you liked their stories. I have tried to faithfully tell their stories to the best of my ability. It is my fervent hope that you will have a better sense of the future you will live in by knowing the wonderful stories of the people who helped to make it possible.

And now, as Mr. Spock of *Star Trek* said, "May you live long and prosper."

George J. Whalen

Appendix A

The information immediately following is from an informative official document of the Federal Communications Commission of the U.S. Government, available as a free download from the FCC web site, www.fcc.gov. It has been included here for the reader's convenience.

 Consumer Guide

Wireless Devices and Health Concerns

Many federal agencies have considered the important issue of determining safe levels of exposure to radiofrequency (RF) energy. In addition to the Federal Communications Commission, federal health and safety agencies such as the Environmental Protection Agency (EPA), the Food and Drug Administration (FDA), the National Institute for Occupational Safety and Health (NIOSH) and the Occupational Safety and Health Administration (OSHA) have been actively involved in monitoring and investigating issues related to RF exposure. For example, the FDA has issued guidelines for safe RF emission levels from microwave ovens, has reviewed scientific literature of relevance to RF exposure (see fda.gov/media/135043/download), and continues to monitor exposure issues related to the use of certain RF devices such as cell phones. Likewise, NIOSH conducts investigations and health hazard assessments related to occupational RF exposure.

Federal, state and local government agencies and other organizations have generally relied on RF exposure standards developed by expert non-governmental organizations such as the Institute of Electrical and Electronics Engineers (IEEE) and the National Council on Radiation Protection and Measurements (NCRP).

Since 1996, the FCC has required that all wireless communications devices sold in the United States meet its minimum guidelines for safe human exposure to radiofrequency (RF) energy. The FCC's guidelines and rules regarding RF exposure are based upon standards developed by IEEE and NCRP and input from other federal agencies, such as those listed above.

For wireless devices intended for use near or against the body (such as cell phones, tablets and other portable devices) operating at or below 6 GHz, these guidelines specify exposure limits in terms of Specific Absorption Rate (SAR). The SAR is a measure of the rate that RF energy is absorbed by the body. For exposure to RF energy from wireless devices, the allowable FCC SAR limit is 1.6 watts per kilogram (W/kg), as averaged over one gram of tissue.

For wireless devices operating in the frequency range above 6 GHz, the guidelines specify power density as the relevant RF exposure limit. Power density is defined as an amount of RF power per unit area. Existing power density limits apply for whole-body exposure, but power density limits for localized exposure are being considered (see the Notice of Proposed Rulemaking in ET Docket No. 19-226, FCC 19-126).

All wireless devices sold in the US go through a formal FCC approval process to ensure that they do not exceed the exposure limits when operating at the device's highest possible power level. If the FCC learns that a device does not conform with the test report upon which FCC approval is based – in essence, if the device in stores is not the device the FCC approved – the FCC can withdraw its approval and pursue enforcement action against the appropriate party. For more information on device testing and SAR for cell phones, go to fcc.gov/consumers/guides/specific-absorption-rate-sar-cell-phnes-what-it-means-you.

FEDERAL COMMUNICATIONS COMMISSION CONSUMER AND GOVERNMENTAL AFFAIRS BUREAU 45 I STREET NE WASHINGTON, DC 20554

Several US government agencies and international organizations work cooperatively to monitor research on the health effects of RF exposure. According to the FDA and the World Health Organization (WHO), among other organizations, to date, there is no consistent or credible scientific evidence of health problems caused by the exposure to radio frequency energy emitted by cell phones. The FDA further states that "the weight of the scientific evidence does not support an increase in health risks from radio frequency exposure from cell phone use at or below the radio frequency exposure limits set by the FCC" (see fda.gov/radiation-emitting-products/cell-phones/scientific-evidence-cell-phone-safety).

The FDA maintains a website on RF issues at fda.gov/Radiation-EmittingProducts/RadiationEmittingProductsandProcedures/HomeBusinessandEntertainment/CellPhones/default.htm.

The WHO has established an International Electromagnetic Fields Project (IEFP) to provide information on health risks, determine research needs and supports efforts to harmonize RF exposure standards. The WHO provides additional information on RF exposure and mobile phone use at who.int/mediacentre/factsheets/fs193/en/index.html. For more information on the IEFP, go to who.int/peh-emf/en.

Some health and safety interest groups have interpreted certain reports to suggest that wireless device use may be linked to cancer and other illnesses, posing potentially greater risks for children than adults. While these assertions have gained increased public attention, currently no scientific evidence establishes a causal link between wireless device use and cancer or other illnesses. Those evaluating the potential risks of using wireless devices agree that more and longer-term studies should explore whether there is a better basis for RF safety standards than is currently used. The FCC closely monitors all of these study results. However, at this time, there is no basis on which to establish a different safety threshold than our current requirements.

You can find additional useful information on the FCC's website at fcc.gov/rfsafety and links to some of the other responsible organizations at fcc.gov/engineering-technology/electromagnetic-compatibility-division/radio-frequency-safety/faq/rf-safety#Q28.

What You Can Do

Even though no scientific evidence currently establishes a definitive link between wireless device use and cancer or other illnesses, and even though all such devices must meet established federal standards for exposure to RF energy, some consumers are skeptical of the science and/or the analysis that underlies the FCC's RF exposure guidelines. Accordingly, some parties recommend taking measures to further reduce exposure to RF energy. **The FCC does not endorse the need for these practices,** but provides information on some simple steps that you can take to reduce your exposure to RF energy from cell phones. **For example,** wireless devices typically emit more RF energy when you are using them. The closer the wireless device is to your body, the more energy you will absorb.

Some measures to reduce your RF exposure include:

- Reduce the amount of time spent using your wireless device.
- Use a speakerphone, earpiece or headset to reduce proximity to the head (and thus head exposure). While wired earpieces may conduct some energy to the head and wireless earpieces also emit a small amount of RF energy, both wired and wireless earpieces remove the greatest source of RF energy (the cell phone or handheld device) from proximity to the head and thus can greatly reduce total exposure to the head.
- Increase the distance between wireless devices and your body.
- Consider texting rather than talking—**but don't text while you are driving**.

Some parties recommend that you consider the reported SAR value of wireless devices. However, comparing the SAR of different devices may be misleading. First, the actual SAR varies considerably depending upon the conditions of use. In particular, while cell phones are tested at their maximum power levels to ensure safety under even the most severe operating conditions, they will typically operate at much lower power levels resulting in RF exposures much lower than the reported SAR values. Cell phones constantly vary their power to operate at the minimum power necessary for communications; operation at maximum power occurs infrequently. Second, the reported highest SAR values of wireless devices do not necessarily indicate that a user is exposed to more or less RF energy from one cell phone than from another during normal use (see our guide on SAR and cell phones at fcc.gov/guides/specific-absorption-rate-sar-cell-phones-what-it-means-you). Third, the variation in SAR from one mobile device to the next is relatively small compared to the reduction that can be achieved by the measures described above.

Consumers should remember that all wireless devices are certified to meet the FCC's maximum SAR limits. These limits incorporate a considerable safety margin. Information about the maximum SAR value for each phone is publicly available on the FCC website at fcc.gov/general/specific-absorption-rate-sar-cellular-telephones, and may be provided with device documentation or by dialing *#07# on certain models. Additional guidance on reducing RF exposure from cell phones is available on the FDA website at fda.gov/radiation-emitting-products/cell-phones/reducing-radio-frequency-exposure-cell-phones.

Other Risks

While current research indicates that cell phones do not seem to pose a significant health problem for pacemaker wearers, some studies have shown that wireless devices might interfere with implanted cardiac pacemakers if used within eight inches of the pacemaker. Pacemaker wearers may want to avoid placing or using a wireless device this close to their pacemaker. Additional information on potential cell phone interference with pacemakers and other medical devices is available on the FDA website at fda.gov/radiation-emitting-products/cell-phones/potential-cell-phone-interference-pacemakers-and-other-medical-devices.

Consumer Help Center

For more information on consumer issues, visit the FCC's Consumer Help Center at_fcc.gov/consumers.

FCC DOCUMENT

Alternate formats

To request this article in an alternate format—braille, large print, Word or text document or audio—write or call us at the address or phone number at the bottom of the page, or send an email to fcc504@fcc.gov.[1]

---

[1] FEDERAL COMMUNICATIONS COMMISSION CONSUMER AND GOVERNMENTAL AFFAIRS BUREAU 45I STREET NE WASHINGTON, DC 20554

About the Author

George J. Whalen is an engineer, marketer, lifelong amateur radio hobbyist (NY9A), and prolific nonfiction writer in technology. He holds a number of patents, has published a great many electronics and tech articles and more than twenty-six books, including *Automotive Electronics, Solid State Ignition, Electrical Wiring, Popular Science How It Works: Illustrated,* and more. He studied at NY Tech and Brooklyn Poly, graduated with a B.S. (with distinction) in 1984 from Iona College, specialized in electronic technology marketing, was a senior executive at a well-known Fortune 100 firm, then opened his own tech marketing consultancy and served many U.S. and international electronic technology clients for over three decades. He's always been an avid student of technology history, inventors and their little-known stories, and how their technologies converge over time to create super-solutions like 5G. His next book is *Automotive Electronics: 1970* and will chronicle its start-up. As Whalen says, "To see how far we've come, you must know how it began."

www.ingramcontent.com/pod-product-compliance
Lightning Source LLC
Chambersburg PA
CBHW020436220526
45464CB00002B/724